Mercury and the Everglades. A Synthesis and Model for Complex Ecosystem Restoration

Curtis D. Pollman • Darren G. Rumbold
Donald M. Axelrad
Editors

Mercury and the Everglades. A Synthesis and Model for Complex Ecosystem Restoration

Volume I – The Evolution of the Everglades as a Perturbed Ecosystem and the Role of Atmospheric Mercury

Springer

Editors
Curtis D. Pollman
Aqua Lux Lucis, Inc
Gainesville, FL, USA

Donald M. Axelrad
College of Pharmacy, Institute of Public Health
Florida A&M University
Tallahassee, FL, USA

Darren G. Rumbold
Department of Marine and Environmental Sciences
Florida Gulf Coast University
Fort Myers, FL, USA

ISBN 978-3-030-20069-5 ISBN 978-3-030-20070-1 (eBook)
https://doi.org/10.1007/978-3-030-20070-1

This Springer imprint is published by the registered company Springer Nature Switzerland AG.
The registered company address is: Gewerbestrasse 11, 6330 Cham, Switzerland

Conservation 5, Everglades National Park. (Photo by Clyde Butcher)

Preface Volume I

The Florida Everglades is an iconic ecosystem. By virtue of its broad expanse alone - the Everglades Protection Area is the largest subtropical wetland in the United States (ca. 960,000 ha) - it is one of the world's most important freshwater wetlands. A key remnant of the historical Everglades, the Everglades National Park (ca. 560,000 ha), has had its ecological significance recognized by its designation as an International Biosphere Reserve, a World Heritage Site, a Ramsar Wetland of International Importance, and a specially protected area under the Cartagena Treaty. It was thus fitting that the Everglades National Park was the first US national park created primarily because of its diversity of plants and animals and its high wildlife population numbers – notably wading birds – rather than because of its scenic values (NPS 2015). The park's usage matches the International Union for Conservation of Nature (IUCN) national park definition "To protect natural biodiversity along with its underlying ecological structure and supporting environmental processes, and to promote education and recreation" (IUCN 2019). The Everglades offers visitors myriad opportunities for passively observing wildlife, including the 68 federally listed species that occur within the greater Everglades (Chap. 1, this volume), while also providing good prospects for recreational taking of fish and wildlife.

In spite of the recognition of its ecological importance, the Everglades is being subjected to multiple critical threats. Decades of progressive encroachment and various modifications to the Everglades leave us currently with less than half the area of the historic ecosystem (SFWMD 2019) through habitat loss to agriculture and urbanization, and yet threats to the Everglades continue. Beyond habitat loss, threats to the Everglades include large-scale hydrologic alterations that have both reduced water availability to this wetland and perturbed the natural cycle of inundation and drying, the influx of a host of invasive plant and animal species, climate change with resultant sea level rise, and water quality. To date, the management of these various threats has focused primarily on restoring water flow and reducing inputs of the plant nutrient phosphorus. Indeed, the state and federal Comprehensive Everglades Restoration Plan is the world's most ambitious ecosystem restoration project with a cost of more than $10.5 billion (NPS 2018).

Less well recognized as a threat to human use of, and wildlife populations in the Everglades, is exposure to the toxic heavy metal, mercury (Hg). In July 1989, one of the approximately 50 Florida panthers (*Felix concolor caryi*) surviving in the wild (Roelke 1990) was found dead in the Everglades under circumstances where the cause of death was not immediately obvious. Subsequent analyses revealed that the female panther contained extremely high concentrations of Hg, leading the researchers to conclude that Hg toxicity was the cause of death (Roelke et al. 1991). Coming on the heels of monitoring data that also showed highly elevated concentrations of Hg in piscivorous fish in both the Everglades and elsewhere in Florida (Ware et al. 1990), the death of the panther was a clarion call to action about a Hg contamination problem threatening the Everglades.

This call to action on the Everglades Hg problem followed a recommendation made two decades earlier in a study by Ogden et al. (1974). Between 1971 and 1973, Ogden et al. (1974) collected biota from the Everglades National Park to assess the contaminant levels and determined Hg levels in fish, frogs, crayfish, alligator eggs, and birds. Based on their results, Ogden et al. (1974) remarked rather presciently that "[T]he relatively high (Hg) concentrations in some freshwater vertebrates deserves further monitoring and study, with emphasis on sources of the mercury…." Unfortunately, the number of samples included in the Ogden et al. (1974) study was rather low. Moreover, there were then neither well-accepted human health-based Hg standards for safe consumption of fish and wildlife nor Hg in tissue standards to assess Hg threat to populations of biota. As a result, both the significance and recommendations for "further (Hg) monitoring and study" that emerged from the Ogden et al. (1974) study were essentially ignored and lay dormant. It remained for an issue concerning a different toxic heavy metal in an area of Florida far from the Everglades, to catalyze the unveiling of the existence of a Hg problem that holds profound ramifications for the Everglades.

Heavy metals monitoring in Florida's surface waters began in earnest in 1982 because of a potential contamination issue regarding the heavy metal lead, when the then Florida Game and Fresh Water Fish Commission (FGFWFC) commenced a survey of metals in fish in the Chipola River in Northwest Florida.

This metals survey was initiated based on the possibility that runoff from a site near the Chipola River where between 1970 and 1981, the Sapp Battery Company had reclaimed lead scrap from lead-acid automobile batteries (HRS 1995), was contaminating the river with lead. The Sapp Battery Company, which processed 50,000 batteries per week at its peak, disposed of used battery acid by pouring it on the ground outside the plant, from where it drained into an adjacent cypress swamp, potentially leading to contamination of the Chipola River. The Sapp Battery site eventually was listed as one of Florida's early superfund sites (Ware et al. 1990).

As part of the Sapp Battery site study, a "control" or background sampling site was established in the Santa Fe River in Northeast Florida to provide comparative data from a similar environment with no point-source industrial pollutant inputs. Fish samples were collected in 1983 by the then Florida Department of Environmental Regulation (FDER), and fish were analyzed for selected pesticides and heavy metals (Bigler et al. 1985).

Hg results for Santa Fe River largemouth bass (*Micropterus salmoides*) were surprisingly high. As an initial response to these unexpected findings, the state's chemical laboratories first verified their analytical techniques and then resampled largemouth bass from the Santa Fe River the following year (1984). The resampling verified the high Hg results obtained during the initial background site sampling and prompted the formation of an interagency task force composed of staff from the FGFWFC, the FDER, and the then Florida Department of Health and Rehabilitative Services (HRS).

A statewide Hg investigation was subsequently initiated that included sampling of about 20 Florida waterbodies annually. In 1988, this ongoing investigation revealed not only elevated concentrations of Hg in largemouth bass and other species collected in the Everglades (Brim et al. 1994), but also that the Hg problem was ubiquitous in Florida and the magnitude troubling. Of the 80 waterbodies sampled, 51 had Hg concentrations in fish approaching or exceeding the Florida health standard of 0.5 mg/kg[1]. The highest Hg concentrations detected were in fish collected from the Everglades Water Conservation Area 3, with mean values in largemouth bass exceeding the HRS limited fish-consumption health standard (0.5 mg/kg) by a factor of over 4. As a result, the first ever public health advisories concerning Hg in fish in Florida were issued, recommending no consumption of any largemouth bass from the Everglades. Limited fish-consumption advisories were issued for an additional 20 waterbodies across the state where mean Hg concentrations in fish ranged from 0.5 to 1.5 mg/kg.

The issuance of public health advisories for fish consumption served as the formal catalyst for organized Florida state and federal interagency investigations into the Everglades Hg problem and led to the creation of a statewide mercury coordinator position in 1992. Dr. Thomas D. Atkeson was appointed to FDER as the first and only mercury coordinator[2] and shortly thereafter organized and oversaw a multiagency research effort – the South Florida Mercury Science Program – consisting of a partnership of federal, state, and local government agencies, academic and private research institutions, and the electric power industry. Larry E. Fink, who about that time joined the South Florida Water Management District (SFWMD), paired with Dr. Atkeson and coordinated Everglades Hg research and monitoring from the district while obtaining funds to conduct research. A third key contributor was Dr. Aaron Higer who in 1992 became US Geological Survey's (USGS) liaison to the SFWMD, FDER, and other state and federal departments regarding Everglades mercury and restoration. This partnership greatly advanced the understanding of the Everglades Hg problem (Atkeson and Parks 2001; Fink et al. 1999).

---

[1] In 1989, Florida HRS established a health standard for total mercury in edible fillets of freshwater fish for fish consumption by adults, with separate standards for women of childbearing age and for children. For mercury in fish ranging from 0.5 mg/kg to 1.5 mg/kg, limited consumption of fish was recommended, while for fish above 1.5 mg/kg mercury, "should not be consumed" was the recommendation (HRS, 1989).

[2] The Florida Department of Environmental Protection (FDEP), which succeeded the FDER in 1993, abandoned the mercury coordinator position in 2010.

At this time, the Everglades Hg issue gained further impetus because the federal government had sued the State of Florida in 1988 over alleged violations of state water quality standards, particularly phosphorus. Stemming from the lawsuit, Florida passed The Everglades Forever Act in 1994 which required the state to "Pursue comprehensive and innovative solutions to the issues of water quality…." This language applied to Hg as well as other water quality issues.

The inception of the South Florida Mercury Science Program (SFMSP) in 1995 encompassed the US Geological Survey's (USGS) Placed-Based Studies Program, a multi-agency federal and state study coordinated by Dr. Higer (Gerould and Higer 1999). The program included the Aquatic Cycling of Mercury in the Everglades (ACME) Project, which was an investigation of Hg biogeochemistry with particular emphasis on inorganic Hg methylation (Stober et al. 1996). Closely linked to this work was landscape-scale monitoring of the Everglades conducted between 1995 and 2014 through the US Environmental Protection Agency (USEPA) with the support of the local cooperators to evaluate the health of the ecosystem status and extent of contamination. Known as the Regional Environmental Monitoring and Assessment Program (R-EMAP), it was implemented in the Everglades with a special focus on quantifying the spatial extent and variations (both spatial and temporal) of the Hg problem with a view to help inform management decisions related to ecosystem mitigation and restoration (Scheidt and Kalla 2007).

Other interagency research included the Florida Atmospheric Mercury Study (FAMS), which was arguably the first regional scale study of Hg deposition in rainfall in the United States, and included six sites sampled between 1993 and 1996 in south peninsular Florida and funded by the FDER, the Electric Power Research Institute (EPRI), the Florida Electric Power Coordinating Group (FCG), and the SFWMD (Pollman et al. 1995, Landing et al. 1995; Gill et al. 1995; and Guentzel et al. 2001). The FDER and USEPA also funded the South Florida Atmospheric Mercury Monitoring Study (SoFAMMS) to evaluate near-source mercury emission impacts on the south Florida urban fringe and the Everglades (Dvonch et al. 1999). The FDER, FCG, and EPRI also collaborated on funding the development of current and historical emissions inventories for Hg in south Florida to further elucidate likely sources impacting the Everglades. Ultimately, atmospheric Hg deposition research lead to the conclusion that > 95% of all external inputs of Hg to the Everglades were atmospheric (USEPA 1996).

Collaborative funding and cooperation from the FDER, USEPA, and EPRI also included extending the Mercury Cycling Model (MCM) to the Everglades. The MCM had originally been developed under the auspices of EPRI to evaluate the primary variables and processes governing Hg biogeochemical cycling in temperate lakes, but required modification so that it could more properly include key wetland processes that are important in the Everglades.

Ultimately, the MCM was employed for use in a pilot total maximum daily limit (TMDL) study funded by the USEPA and conducted through the Florida Department of Environmental Protection (FDEP). The pilot TMDL defined the maximum loading of mercury to the ecosystem that would be protective of human health, where the exposure was via the consumption of largemouth bass (Atkeson et al. 2003). A main conclusion of this pilot TMDL was that "In the absence of changes to the system other than mercury loading that affects the cycling of mercury (e.g. changes in sulfur cycling, nutrient cycling, or hydrology), a reduction of about 80% of current annual mercury deposition rates would be needed for the mercury concentrations in a 3-year old largemouth bass to be less than Florida's present fish consumption advisory action level of 0.5 ppm" (mg/kg).

At this time, research indicated that atmospheric Hg emissions from South Florida had been greatly reduced since the early 1990s and that atmospheric Hg sources external to Florida along with atmospheric chemistry and South Florida meteorology could be the cause of the elevated atmospheric deposition of Hg to the Everglades (Chap. 6, this volume). In part because of this realization, research on Everglades sulfur accelerated in order to examine if Everglades methylmercury – a highly bioaccumulative organic Hg species – could be adequately mitigated via biogeochemical control of Hg methylation by naturally-occurring sulfate-reducing bacteria through limiting sulfur inputs to the ecosystem (Bates et al. 2002; Axelrad et al. 2007; Axelrad et al. 2013; Orem et al. 2011). Unfortunately, this line of research is contentious and remains incomplete (Orem et al. 2011). Elaboration on the sulfur and Hg issue is a key topic of Volume II of this book.

That above is but an abbreviated listing of the full breadth of research conducted on the Hg problem that followed Dr. Atkeson's appointment as the FDER mercury coordinator. It is thus perhaps one of the great ironies of our awakening environmental awareness that the Everglades and the Hg problem in particular is one of the more prolifically studied environmental problems in the world, both in terms of studies across time and the amount of resources expended, and yet policy-makers have done little to attempt to develop an integrated approach toward understanding and resolving the problem. The broad expanse and wealth of this research have never been fully synthesized and integrated into a framework so that the policy-makers could begin the process of making informed decisions.

The objective of this book is to provide such a synthesis and framework. And, while the advantages of such an effort to the Everglades are obvious, the benefits of this integrative effort extend globally and well beyond its boundaries and immediate environs. Among these are:

1. The Everglades is a paradigm for aquatic ecosystem response and potential restoration strategies. Included in this paradigm are the complexities that are inherent in many aquatic ecosystems, including nonlinear response-driver relationships, and the risk that certain control strategies could exacerbate the problem if not properly understood and implemented.
2. Consistent with the first benefit, and because the Everglades is perhaps the world's most studied aquatic ecosystem with respect to the Hg problem, the results of a comprehensive synthesis of that research can help direct future research on other aquatic ecosystems and elucidate the questions related to policy toward those gaps that are truly the most critical.
3. The solution to the Everglades Hg problem as proposed by the State of Florida (FDEP 2013) and accepted by the USEPA will not result in toxicologically acceptable Hg levels in fish for many decades, and will therefore not make fish

safe for human or wildlife consumption for that term. As a result, it is worthwhile to reexamine Everglades' Hg science and evaluate the likely efficacy of solutions in addition to those of the state and federal governments.

This book, *Mercury and the Everglades: A Synthesis and Model for Complex Ecosystem Restoration* is organized into three volumes:

Volume I – *The Evolution of the Everglades as a Perturbed Ecosystem and the Role of Atmospheric Mercury*
Volume II – *Aquatic Mercury Cycling and Bioaccumulation in the Everglades*
Volume III – *Temporal Trends of Mercury in the Everglades, Synthesis and Management Implications*

Each volume of this book is organized into separate parts that include a series of individual chapters aligned with the part theme. These parts include:

Part 1 "Introduction to the Everglades" (Volume I)
Part 2 "Atmospheric Mercury" (Volume I)
Part 3 "Aquatic Mercury" (Volume II)
Part 4 "Mercury Bioaccumulation" (Volume II)
Part 5 "Temporal Changes in the Mercury Signal in the Everglades" (Volume III)
Part 6 "Integration: Towards a Unified Model of Mercury Cycling in the Everglades (Volume III)

Parts 2 through 4, which focus on specific aspects of the Hg issue, each include an introductory chapter and conclude with synthesis chapters integrating key aspects of each individual chapter as they relate to the Everglades Hg problem. Part 5 also concludes with a synthesis chapter. The book itself concludes with Part 6 which includes two chapters devoted to distilling key elements of the extant Hg science into a management model framework ultimately accessible to policy- and decision-makers charged with protecting the Everglades, and a final chapter that is used by the editors to share with the reader lessons we have learned undertaking the process of compiling this book.

Volume I includes six chapters:

Chapter 1, "Overview of the Everglades," spans the evolution, geology, climate, plant and wildlife communities, modifications, and restoration.
Chapter 2, "The Evolution of the Everglades as a Perturbed Ecosystem," covers the natural system and anthropogenic modifications, including altered hydrology and water quality.
Chapter 3, "Overview of the Atmospheric Mercury Cycle," discusses the present sources of atmospheric deposition of mercury, importance of mercury speciation on transport and deposition, and factors influencing mercury deposition to the Everglades.
Chapter 4, "Atmospheric Deposition Flux of Mercury to the Everglades," describes mercury wet deposition and dry deposition to the Everglades and the fact that the atmospheric deposition flux of mercury in Southern Florida has been among the highest in the United States over the past two decades.

Chapter 5, "Mercury Emission Sources and Contributions of Atmospheric Deposition to the Everglades," presents a summary of the literature on mercury air emission sources in South Florida, the remainder of Florida, the United States, and the world, along with their historical and current contributions to atmospheric deposition of mercury to the Everglades.

Chapter 6, "Atmospheric Deposition of Mercury in the Everglades: Synthesis of Global Cycling and Local Process Effects," synthesizes information on global cycling of mercury in the atmosphere with information on atmospheric deposition dynamics of mercury in the Everglades and the relative importance of different atmospheric mercury sources to the Everglades.

We trust that this volume and its companions will provide readers with insights regarding the Everglades mercury issue and leads on how to approach mitigating mercury in other ecosystems.

Tallahassee, FL, USA — Donald M. Axelrad
Gainesville, FL, USA — Curtis D. Pollman
Fort Myers, FL, USA — Darren G. Rumbold

References

Atkeson T, Parks P (2001) Everglades consolidated report Chapter 7: the Everglades mercury problem. South Florida Water Management District, West Palm Beach, p 41. http://my.sfwmd.gov/portal/page/portal/pg_grp_sfwmd_sfer/portlet_prevreport/consolidated_01/chapter%2007/ch07.pdf

Axelrad DM, Atkeson TD, Lange T, Pollman CD, Gilmour CC, Orem WH, Mendelssohn IA, Frererick PC, Krabbenhoft DP, Aiken GR, Rumbold DG, Scheidt DJ, Kalla PI (2007) Mercury monitoring, research and environmental assessment in South Florida. South Florida Environmental Report, Vol. 1, Chapter 3B, South Florida Water Management District, 57 pp https://my.sfwmd.gov/portal/page/portal/pg_grp_sfwmd_sfer/portlet_prevreport/volume1/chapters/v1_ch_3b.pdf

Axelrad DM, Pollman C, Gu B, Lange T (2013) Mercury and Sulfur Environmental Assessment for the Everglades. South Florida Environmental Report, vol. 1, Chapter 3B, South Florida Water Management District, West Palm Beach, FL, 57 pp. http://my.sfwmd.gov/portal/page/portal/pg_grp_sfwmd_sfer/portlet_prevreport/2013_sfer/v1/chapters/v1_ch3b.pdf

Bates AL, Orem WH, Harvey JW, Spiker EC (2002) Tracing Sources of Sulfur in the Florida Everglades. J. of Environ. Qual. 31:287–299

Bigler WJ, Ware F, Savage T, King S, Hartwig C (1985) Heavy metals in fish and clams from the Chipola and Santa Fe Rivers of north Florida. Fla. Acad. Sci. 9pp.

Brim MS, Bateman DH, Jarvis RB, Carmody GA (1994) Mercury in largemouth bass and spotted gar of the Florida Panther National Wildlife Refuge. Publication no. PCFO·EC 94-04. U.S. Fish and Wildlife Service. Ecological Services. 28 pp. https://ecos.fws.gov/ServCat/DownloadFile/21776?Reference=23257

Dvonch JT, Graney JR, Keeler GJ, Stevens RK (1999) Use of elemental tracers to source apportion mercury in South Florida precipitation. Environ. Sci. and Technol. 33:4522–4527

FDEP (2013) (Florida Department of Environmental Protection). Mercury TMDL for the State of Florida. Final report. Watershed evaluation and TMDL section. October 24, 2013. https://floridadep.gov/sites/default/files/Mercury-TMDL.pdf

Fink L, Rumbold D, Rawlik P (1999) Everglades interim report Chapter 7: The Everglades mercury problem. South Florida Water Management District, West Palm Beach FL. 58 pp. http://apps.sfwmd.gov/sfwmd/SFER/1999_Everglades_Interim_Report/ever_interim_rep_1999.pdf

Gerould S. and Compilers AH (1999) U.S. geological survey. Open-File Report 99-181 U.S. geological survey program on the South Florida ecosystem. In: Proceedings of South Florida restoration science forum, May 17–19, 1999, Boca Raton, Florida, 132 pp. https://fl.water.usgs.gov/PDF_files/ofr99_181_gerould.pdf

Gill GA, Guentzel JL, Landing WM, Pollman CD (1995) Total gaseous mercury measurements in Florida: The FAMS project (1992–1994). Water, Air, and Soil Pollution 80:235–244

Guentzel JL, Landing WM, Gill GA, Pollman CD (2001) Processes Influencing Rainfall Deposition of Mercury in Florida. Environ. Sci. Technol. 35:863–873

HRS (1989) Florida Department of Health and Rehabilitative Services. Exposure Guidelines for Mercury in Fish. Inter-Office Memorandum, March 3, 1989. To: Charles S. Mahan, M.D., Deputy Secretary for Health and State Health Officer. From: Richard W. Freeman, Ph.D. Director, Toxicology and Hazard Assessment. 1 p.

HRS (1995) Florida Department of Health and Rehabilitative Services. Health Consultation. Sapp Battery and Salvage Superfund Site. Jackson County, Florida. CERCLIS No. FLD980602882. September 28, 1995. Prepared under a Cooperative Agreement with the Agency for Toxic Substances and Disease Registry, U.S. Public Health Service, Department of Health and Human Services. 18 pp. http://www.floridahealth.gov/environmental-health/hazardous-waste-sites/_documents/s/sappbatterysalvage092895.pdf

IUCN (2019) (International Union for Conservation of Nature). Protected area categories. Accessed Jan. 29, 2019. https://www.iucn.org/theme/protected-areas/about/protected-area-categories

Landing W, Perry JJ, Guentzel JL, Gill GA, Pollman CD (1995) Relationships between the atmospheric deposition of trace elements, major ions, and mercury in Florida: The FAMS project (1992–1993). Water, Air and Soil Poll. 80:343–352

NPS (National Parks Service) (2015) Everglades Natural Features & Ecosystems. https://www.nps.gov/ever/learn/nature/naturalfeaturesandecosystems.htm

NPS (National Parks Service) (2018) Comprehensive Everglades Restoration Plan (CERP). https://www.nps.gov/ever/learn/nature/cerp.htm

Ogden, JC, Robertson WB, Davis GE, Schmidt TW (1974) Pesticides, polychlorinated biphenyls and heavy metals in upper food chain levels, Everglades National Park and Vicinity: National Park Service PB-231 359, 27 pp.

Orem W, Gilmour C, Axelrad D, Krabbenhoft D, Scheidt D, Kalla P, McCormick P, Gabriel M, Aiken G (2011) Sulfur in the South Florida Ecosystem: Distribution, Sources, Biogeochemistry, Impacts, and Management for Restoration. Crit Rev Environ Sci Technol. 41(6):249–288. http://www.informaworld.com/smpp/content~db=all?content=10.1080/10643389.2010.531201

Pollman CD, Gill GA, Landing Guentzel WM, Bare D, Porcella D, Zillioux E, Atkeson T (1995) Overview of the Florida Atmospheric Mercury Study (FAMS). Water Air Soil Pollut. 80:285–290

SFWMD (2019) (South Florida Water Management District) Everglades. https://www.sfwmd.gov/our-work/everglades

Stober J, Scheidt D, Jones R, Thornton K, Ambrose R, France D (1996) South Florida ecosystem assessment. Monitoring for adaptive management: implications for ecosystem restoration. Interim report. United States Environmental Protection Agency EPA-904-R-96-008. 26 pp. https://www.epa.gov/sites/production/files/2014-03/documents/sfleair.pdf

Roelke ME (1990) Florida panther biomedical investigation: health and reproduction. final performance report. Endangered species Project E-I II-E-6 7506. Florida Game and Freshwater Fish Commission. 176 pp.

Roelke ME, Schultz DP, Facemire CF, Sundlof SF, Royals HE (1991) Mercury contamination in Florida panthers. Report of the Florida Panther Technical Subcommittee to the Florida Panther Interagency Committee. December 1991. 58 pp.

Scheidt DJ, Kalla PI (2007) Everglades ecosystem assessment: water management, water quality, eutrophication, mercury contamination, soils and habitat. In: Monitoring for adaptive management: a R-EMAP status report. EPA 904-R-07-001. United States Environmental Protection Agency, Atlanta. https://www.nps.gov/ever/learn/nature/upload/MON03-4FinalReportSecure.pdf

SFWMD (2019) (South Florida Water Management District). Everglades. https://www.sfwmd.gov/our-work/everglades

USEPA (1996) (United States Environmental Protection Agency). 1996. South Florida Ecosystem Assessment Interim Report. Monitoring for adaptive management: implications for ecosystem restoration. Region 4 SESD and ORD. EPA-904-96-008. December 1996. https://www.epa.gov/sites/production/files/2014-03/documents/sfleair.pdf

Ware FJ, Royals HE, Lange TR (1990) Mercury contamination of Florida largemouth bass. Proc Annu Conf SE Assoc Fish Wild Agencies 44:5–12

Acknowledgments

The explosion of research on the mercury problem in the Everglades that began in the early 1990s came on the heels of a revolution in the science of mercury in natural waters made possible by the advances in sampling and analysis of mercury at ultra-trace concentrations. This revolution was led in part by Dr. William Fitzgerald of the University of Connecticut and his then graduate students Gary Gill and Nicolas Bloom; it was subsequently seized upon and further supported most notably by Dr. Donald Porcella with the Electric Power Research Institute (EPRI). That support led to the Mercury in Temperate Lakes Program which in turn fueled the development and maturation of many key scientists who subsequently comprised much of the initial scientific foundation for mercury research efforts within the Everglades and elsewhere in Florida.

The first decade or so of Everglades mercury research was multidisciplinary and ambitious in scope. As such, it required large amounts of funding and participation of multiple governmental, academic, and private entities, including the State of Florida Department of Environmental Regulation (FDER), the South Florida Water Management District (SFWMD), the US Geological Survey (USGS), the US Environmental Protection Agency (USEPA), the EPRI, and the Florida Electric Power Coordinating Group (FCG). Key to bring these disparate entities to the table and forge a productive alliance were Porcella, Dr. Thomas Atkeson, and Dr. Edward Zillioux. Dr. Atkeson had been recently appointed to the FDER (later reorganized to become the Florida Department of Environmental Protection) as mercury coordinator to help organize and oversee the resulting multi-agency research effort which became known as the South Florida Mercury Science Program. Dr. Zillioux, who served on the Environmental Committee of the FCG, was instrumental in securing financial support from the FCG for mercury-related research in both the Everglades and elsewhere within the state.

Other individuals key to the success of the South Florida Mercury Science Program included Larry E. Fink and Dr. Aaron Higer. Mr. Fink, who about that time had joined the SFWMD, paired with Dr. Atkeson and coordinated Everglades mercury research and monitoring from the SFWMD while obtaining funds to conduct research. In 1992, Dr. Higer became the USGS liaison to the SFWMD, FDER, and other state and federal departments regarding Everglades mercury and restoration. This partnership greatly advanced the understanding of the Everglades mercury problem.

It is rare that any undertaking of substantial worth and time is not marked by the passing of individuals who contributed greatly to the effort. Such is true for the two-plus decades of Everglades mercury research, and we sadly note the losses of Dr. Atkeson, Dr. Porcella, Dr. Higer, Dr. Gerald Keeler (University of Michigan), Dr. George Aiken (USGS), Dr. Carl Miles (University of Florida), and Dr. David Evans (National Oceanic and Atmospheric Administration). The contributions of each were critical to the advancement of Everglades mercury science across multiple disciplines, and we are diminished by their passing.

The necessity of synthesizing the decades of Everglades mercury research had been the topic of idle conversation between two of the coeditors (Pollman and Axelrad) for a number of years but had never progressed beyond that point until Ed Zillioux approached us in 2014 about undertaking the effort of coediting this book. Shortly thereafter, Dr. Zillioux stepped aside because the demands of the book outstripped the physical limitations imposed on him by failing health and the care of his wife. This led to our third coeditor (Dr. Darren Rumbold) agreeing to step into the breach left by Dr. Zillioux's absence. Without Dr. Zillioux's vision and initial efforts to bring coherence to the project, this book would never have moved beyond a mere diaphanous concept, and we gratefully acknowledge his substantial role in its completion.

Lastly, the senior coeditor of Volume I wishes to gratefully acknowledge the financial support from the Bullsugar Alliance and Friends of the Everglades. Funding through both groups was instrumental in partially supporting the writing of Chap. 6, Volume I; Chap. 6, Volume II; and Chaps. 1, 3, 5, 6, 7, Volume III.

Contents

Contributors

Donald M. Axelrad College of Pharmacy, Institute of Public Health, Florida A&M University, Tallahassee, FL, USA

Christopher D. Holmes Department of Earth, Ocean and Atmospheric Science, Florida State University, Tallahassee, FL, USA

William M. Landing Department of Earth, Ocean and Atmospheric Science, Florida State University, Tallahassee, FL, USA

Thomas E. Lodge Consulting Ecologist, Asheville, NC, USA

Curtis D. Pollman Aqua Lux Lucis, Inc, Gainesville, FL, USA

Darren G. Rumbold Department of Marine and Environmental Sciences, Florida Gulf Coast University, Fort Myers, FL, USA

Krish Vijayaraghavan Ramboll US Corporation, Novato, CA, USA

Chapter 1
Overview of the Everglades

Thomas E. Lodge

Abstract The Everglades is a freshwater wetland located in southern Florida. It originated 5500 years ago when rising waters of Lake Okeechobee and its vicinity fostered deposition of wetland soils through former uplands with widely exposed bedrock. Its development occurred in the presence of, and under the influence of Native Americans, their occupation having predated the Everglades by thousands of years. Prior to regional drainage, which began in the 1880s, the Everglades covered about 4000 square miles. It sloped southward at about 2.5 inches per mile from Lake Okeechobee to coastal tidal waters at the south end of Florida's mainland. Just over half of the Everglades remains today, mostly broken into water-conservation impoundments controlled by levees and canals with only Everglades National Park at the south end still free-flowing. Pristine areas of the Everglades are oligotrophic, consisting mostly of marshes rich in periphyton. The marshes and periphyton are rapidly degraded by phosphorus enrichment over the maximum background of 10 ppb. Other enrichment of concern is sulfur and its association with elevated methylmercury, the latter biomagnified in Everglades fauna. Everglades marsh landscapes include long-hydroperiod sloughs (flooded 11+ months annually) and intermediate-hydroperiod sawgrass ridges (flooded 9–10 months), both underlain by peat soils, and smaller areas of short-hydroperiod, mixed-herb marshes (mostly flooded 3–7 months) underlain by marl. The latter occur around the edges of the southern Everglades. Unevenly distributed through these marshes are tree islands of several kinds based on their genesis and flora. One type was frequented through history by Native Americans, which influenced island development. Recent work has demonstrated the role of water flow in the evolution and maintenance of marsh and tree-island landscape features, all aligned with the pre-drainage direction of flow. Other Everglades features include small ponds called alligator holes that have various origins but are maintained by alligators and are ecologically important. Surrounding the Everglades are other plant communities, principally forests, and most outflows of water from the Everglades pass through extensive and highly productive tidal mangrove swamp forests before entering shallow marine waters at the

T. E. Lodge (✉)
Consulting Ecologist, Asheville, NC, USA

C. D. Pollman et al. (eds.), *Mercury and the Everglades. A Synthesis and Model for Complex Ecosystem Restoration*,
https://doi.org/10.1007/978-3-030-20070-1_1

south end of the system. Everglades wildlife responds to summer wet-season and winter-spring dry-season cycling, which characterizes southern Florida's nearly tropical climate. Many introduced plants and animals have stressed the natural Everglades ecosystem, which supports 68 threatened and endangered species. Restoration efforts are in progress. Phosphorus reduction, initiated in the mid-1990s, has been successful but short of compliance targets. Overall ecosystem restoration is ongoing but slow. In combination with phosphorus reduction, it involves revision of South Florida's water-control system to ensure the right quality, timing, and distribution of water.

Keywords Everglades · Lake Okeechobee · Wetland · Marsh · Tree Island · Native American · Hydroperiod · Phosphorus · Flow · Peat · Marl · Restoration

1.1 Introduction – What Is the Everglades?

The Everglades[1] is unique. From inception, the name has designated the contiguous, primarily herbaceous freshwater wetlands of South Florida. Prior to regional drainage, which began in 1882 but not in earnest until 1913, the Everglades extended from the south rim of Lake Okeechobee southward to tidal waters at the Florida peninsula's south end (Fig. 1.1), including the Gulf of Mexico, Florida Bay, and Biscayne Bay. Early English maps labeled the area "River Glades" (Douglas 1947), but an 1823 map (Vignoles 1823) used "Ever" instead of "River," labeling the region "The Ever Glades," the first known use of the name. Florida's statehood map (State of Florida 1845) and the "Ives" military map (Ives 1856) show it as one word, "Everglades." The Ives map was the first to have reasonably clear boundaries of the Everglades, including its edge on the west, much of which bordered what is labeled "Big Cypress Swamp," and its eastern edge where it abutted uplands along Florida's southeast coast. More refined Everglades boundaries based on vegetation were mapped by Harshberger (1913) and Davis (1943), the latter a monumental work that included the use of 1940 aerial photographs (see Fig. 1.5). Other boundaries have referenced the extent of peat soils (Gleason and Stone 1994; Richardson 2008), but the Everglades is a geographic place rooted in history, not a science-based set of characteristics that might apply elsewhere. In that light, the mapped extent of the Everglades shown by Marjory Stoneman Douglas in her popular, landmark book, *The Everglades: River of Grass* (1947), the first published book specifically about the Everglades, is a defining standard. Her boundaries have been refined by McVoy et al. (2011) based on multiple sources including early surveyors' notes and modern mapping tools. These boundaries form the convention accepted by this author as the

[1]The name "Everglades" is used here a singular noun following most technical literature such as Loveless (1959), Davis and Ogden (1994), and McVoy et al. (2011). The plural form was used in an 1848 federal reconnaissance report (U.S. Senate 1911) and is common in popular literature such as Douglas (1947). The author discussed this issue with Ms. Douglas, who agreed, "Singular is appropriate but both are grammatically correct.")

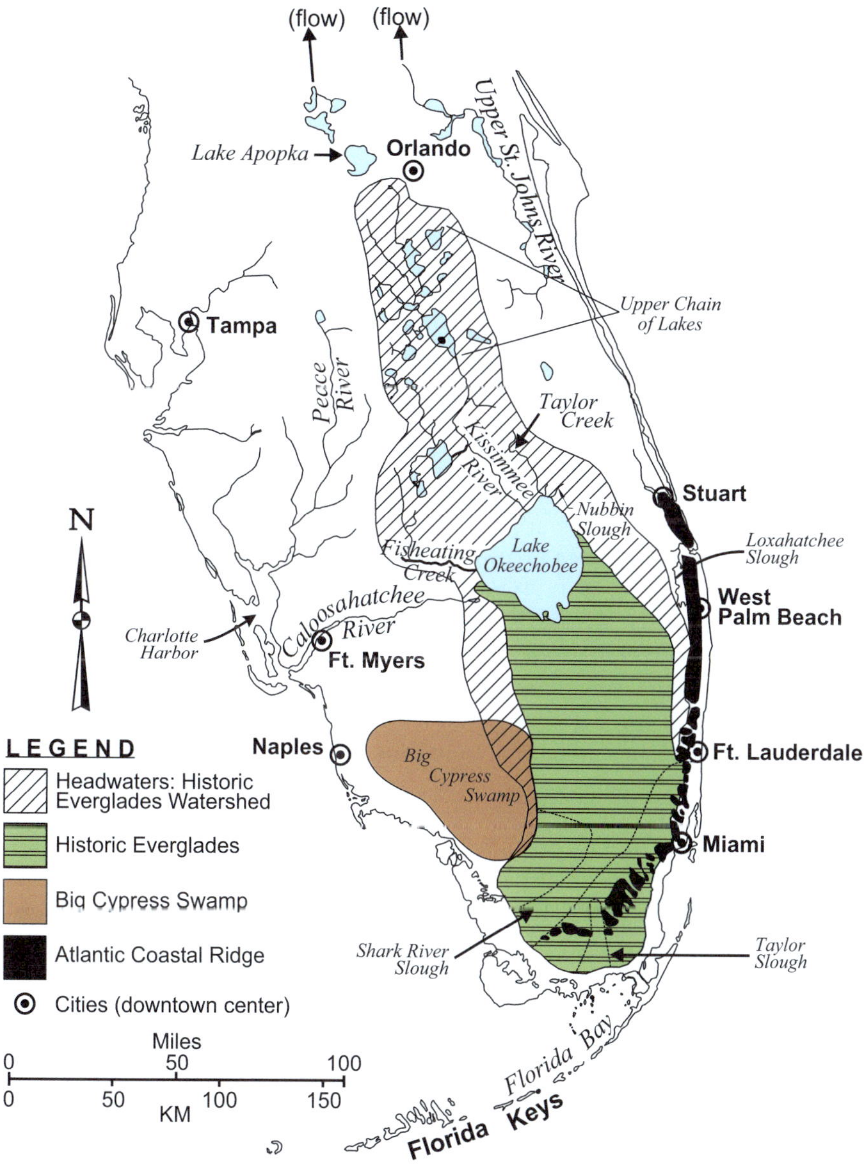

Fig. 1.1 Map of the Florida peninsula showing the historic Everglades and its watershed as described in the text. (Source: Lodge 2017, with permission of Taylor & Francis Group, LLC)

original extent of the Everglades and are further discussed below under "The Predrainage Everglades Landscapes: the Basis for Understanding Change."

The waters of the predrainage Everglades were contiguous with Lake Okeechobee (McVoy et al. 2011; Parker 1984). This hydrologic relationship had been generally known but was first documented in an 1848 reconnaissance report to Congress by Buckingham Smith (U.S. Senate 1911). That report summarized knowledge of the

Everglades and Lake Okeechobee, identified that the source of the lake's water was primarily the Kissimmee River, and confirmed that the lake broadly overflowed into the Everglades across its southern shore. Because of this hydrologic association, the watershed of Lake Okeechobee is included in characterizing the Everglades ecosystem. The Everglades watershed begins in the Kissimmee River's headwaters near Orlando and includes all contributing flows to the lake as well as to the Everglades along its boundary south of Lake Okeechobee (see Fig. 1.1). The names "greater Everglades ecosystem" or "South Florida ecosystem" have been used for the entire watershed (sometimes referenced as the Kissimmee-Lake Okeechobee-Everglades [KLOE] system) together with connected adjacent coastal areas including the Caloosahatchee and St. Lucie Rivers. While the predrainage Everglades overlapped with the headwaters of the Caloosahatchee on the west side of Lake Okeechobee, there was no stream or river connection, just intervening wetlands where modeling indicates that there were only minor flows toward the Caloosahatchee (Said and Brown 2013). On the east side of the lake there was no outlet to coastal waters. Today's St. Lucie Canal, which carries releases eastward from the lake to an ocean outlet via the St. Lucie River at Stuart, Florida, is man-made. As further detailed below, almost all surface outflows from Lake Okeechobee were southward into the Everglades (McVoy et al. 2011; Parker 1984).

Applications of the Everglades name, often shortened to a vernacular, "the Glades," often depart from the geographic convention used here. Lay people commonly lump the Big Cypress Swamp together with the Everglades, and advocacy groups have called the Big Cypress the "Western Everglades." There is also an Everglades restoration component called the "Western Everglades Restoration Project" (USACE 2016) that involves lands in, and north of the Big Cypress Swamp. The largest departure in usage was by the State of Florida in its Northern Everglades and Estuaries Protection Program legislation of 2007 (SFWMD 2008). That program labeled the watersheds of Lake Okeechobee, the Caloosahatchee, and the St. Lucie River as the "Northern Everglades," and defined the "Southern Everglades" being everything south of those areas, including the historic Everglades, the Big Cypress Swamp, and bordering lands. The latter designations are shown in Fig. 1.2 together with a widely accepted extent of the Big Cypress Swamp (McPherson 1984) and Everglades subdivisions, namely "Northern," "Central," and "Southern Everglades" as chosen by the author from numerous scientific literature sources (Gawlik et al. 2002, Goodrick 1984; Heisler et al. 2002; Ogden 1994; Rader 1994; Scheidt and Kalla 2007; USACE 2014) and personal communications. Of the latter subdivisions, the boundary between the northern and central Everglades is not universal. It is shown here conforming to most sources as the modern alignment of Interstate 75 (Alligator Alley), but some sources use the Palm Beach-Broward county line. The boundary between the central and southern Everglades is widely understood to be the Tamiami Trail. In this book, we have endeavored to use "Everglades" in the strict historic sense and its northern, central, and southern portions within that historic boundary as defined above. Locations of observations in other areas, or places that diverge from these definitions are qualified as needed for clarity.

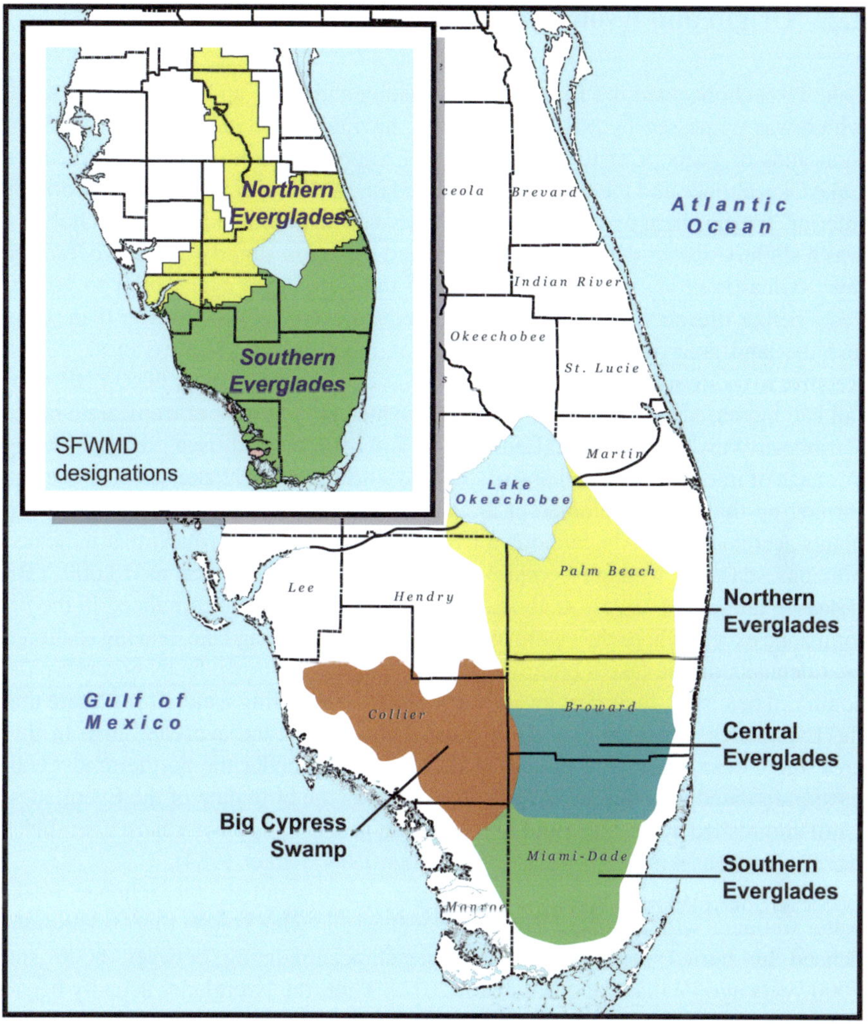

Fig. 1.2 Subdivision of the Everglades: SFWMD designations versus the historic terminology used in this book. (Source: Lodge 2017, with permission of Taylor & Francis Group, LLC)

As a final note on Everglades boundaries, there is no convention defining the Everglades from coastal plant communities at its southern end (unlike the obvious northern boundary on Lake Okeechobee). Douglas (1947) shows the southern coastal areas as being part of the Everglades. Harshberger (1913) and Davis (1943) exclude mangrove swamps and various coastal marshes from the Everglades. McVoy et al. (2011) shows no southern boundary, leaving it open to interpretation. This author defines the Everglades as a freshwater ecosystem and thus excludes saline coastal areas from the Everglades.

1.2 Origin and Evolution of the Everglades

Lake Okeechobee and the Everglades are young and share an origin history, all of which was witnessed by Native Americans, having inhabited southern Florida for thousands of years prior to the existence of these landform features (Carr 2002). Lake Okeechobee and the Everglades evolved in a shallow trough in south Florida's interior. The northern end of the trough – the area that became Lake Okeechobee – was a slightly deeper depression, its minimum elevation slightly below modern sea level (Gleason et al. 1984). At the close of the last glacial cycle, prior to 15,000 years before present (YBP), sea level had been as much as 130 m lower than today and the land area of peninsular Florida was much larger (Wanless et al. 1994). Relative to modern conditions, the climate was cool and dry, but by 6000 YBP, rainfall had increased in a warming trend documented by an influx of tropical plants of Caribbean/West Indian origin (Long 1984), and sea level had risen greatly, slowing drainage of interior lands so that surface waters accumulated. Lake Okeechobee had formed by that time (Steinman et al. 2002). It had a littoral zone where wetland plants accreted peat soils, laid down over earlier calcitic mud (marl) that indicated previous seasonal wetland conditions that had started as much as 13,000 YBP (Gleason and Stone 1994). As the lake level rose, peat soils accumulated in the littoral zone except where they were overwhelmed by the rising lake, leaving scattered peat deposits on the lake's bottom (Kirby et al. 1989). But wetlands bordering the south rim persisted, their peat levels keeping pace with rising water. It was there that the Everglades was born as a "biological dam" against the evolving lake. In that area, the oldest Everglades peat deposits, those deep under the northernmost end, have been dated at about 5500 YBP, often cited as the birth date of the Everglades. Until diminished by drainage and development, peat soils in the northern Everglades were as much as 4 m deep (Gleason and Stone 1994; Parker 1984).

Prior to the evolution of the Everglades associated with Lake Okeechobee, freshwater wetlands with abundant sawgrass developed in what is now Florida Bay, evidenced by buried peat deposits that began accumulating between 6000 and 7000 years ago (Willard and Bernhardt 2011). Thus, the Everglades actually began at both ends of the historic ecosystem. But at the south end, rising sea level overtook those wetlands during the formation of Florida Bay nearly 5000 YBP. Accordingly, marine sediments began accumulating atop the earlier freshwater marl and peat soils (Gleason et al. 1984, Gleason and Stone 1994). In a geologically short time, only a few thousand years, the Everglades was established between Lake Okeechobee and Florida Bay. Prior to drainage modifications, it sloped from a high just over 20 ft at the lake to sea level over a distance of nearly 100 miles, giving an average slope of about 2.5 inches per mile (4 cm/km) (McVoy et al. 2011).

Human presence during the evolution of the Everglades lends an important perspective. Before its existence, Native American habitation in the area to become the Everglades was associated with limestone sinkholes where water was available (Carr 2002). But with the development of the Everglades, widespread peat and peripheral marl soils began covering the interior bedrock trough. The rough, uneven, and widely exposed limestone surface became overlain by these wetland soils, obscuring the bedrock irregularities and filling sinkholes. Intervening wet and dry cycles caused successional changes that especially influenced the evolution of tree islands (Willard and Bernhardt 2011), and episodes of African dust periodically enriched the ecosystem (Glaser et al. 2013). But in its final form, the Everglades became an oligotrophic, phosphorus-limited ecosystem highly sensitive to enrichment above 10 ppb (Gaiser et al. 2005). Through all of these changes, Native Americans adapted, transitioning from the use of rough terrain with limited fresh water to using tree islands in a widely flooded marsh landscape (Ardren et al. 2016; Carr 2002; Graf et al. 2008).

1.3 The Geologic Setting

Looking farther back into geologic history than the Everglades origin summarized above, the Florida peninsula is deeply underlain by an ancient basement ridge of volcanic rock of tectonic origin. In South Florida, the basement ridge is overlain by about 20,000 ft of sedimentary rocks, laid down over the slowly subsiding basement. Below about 1000 ft, these rocks are primarily limestones, dolomites, and evaporates ranging in age from Cretaceous (or perhaps earlier) to Miocene. Atop those sequences, from early Miocene to Holocene, the sediments are a mixture of locally produced carbonates primarily of shallow marine origin, including shells, in addition to clastics of Appalachian origin, including silica sand and clays (Missimer 1984).

Of specific relevance to the Everglades are the geologic formations at or very near the surface, shown in Fig. 1.3. Those forming the floor of the Everglades trough and the Atlantic Coastal Ridge (ACR) accumulated during the last high stand of sea level (Sangamon interglacial period) about 100,000 YBP. Along the east coast, sands and shells of the Anastasia formation were deposited, producing the northern part of the ACR (see Figs. 1.1 and 1.3). This sandy topography continues south where it intergraded into oolitic Miami Limestone laid down in a shallow marine environment as calcium carbonate grains, "ooids." The oolitic limestone, forming the southern extension of the ACR, continued as an elevated outcropping from Miami south and southwest into what is now Everglades National Park where it slopes downward to an end, merging into wetland elevations of the non-oolitic and contemporary western portion Miami Limestone (Gleason and Stone 1994).

The floor of the Everglades trough west of the ACR is composed of two limestone formations. The northern half is the Fort Thompson Formation, which also underlies and surrounds most of Lake Okeechobee. Beneath the lake, the limestone is covered by recently displaced sand deposits related to riverine inputs and redistribution within the lake. The upper portion of the Fort Thompson is a Sangamon interglacial deposit, but the formation continues below as similar but older strata of earlier interglacial periods. It is composed mostly of marine limestone but has interspersed layers with marl and fossils of freshwater origin, indicating varying conditions with changes in sea level. In some locations, especially in the northeastern Everglades,

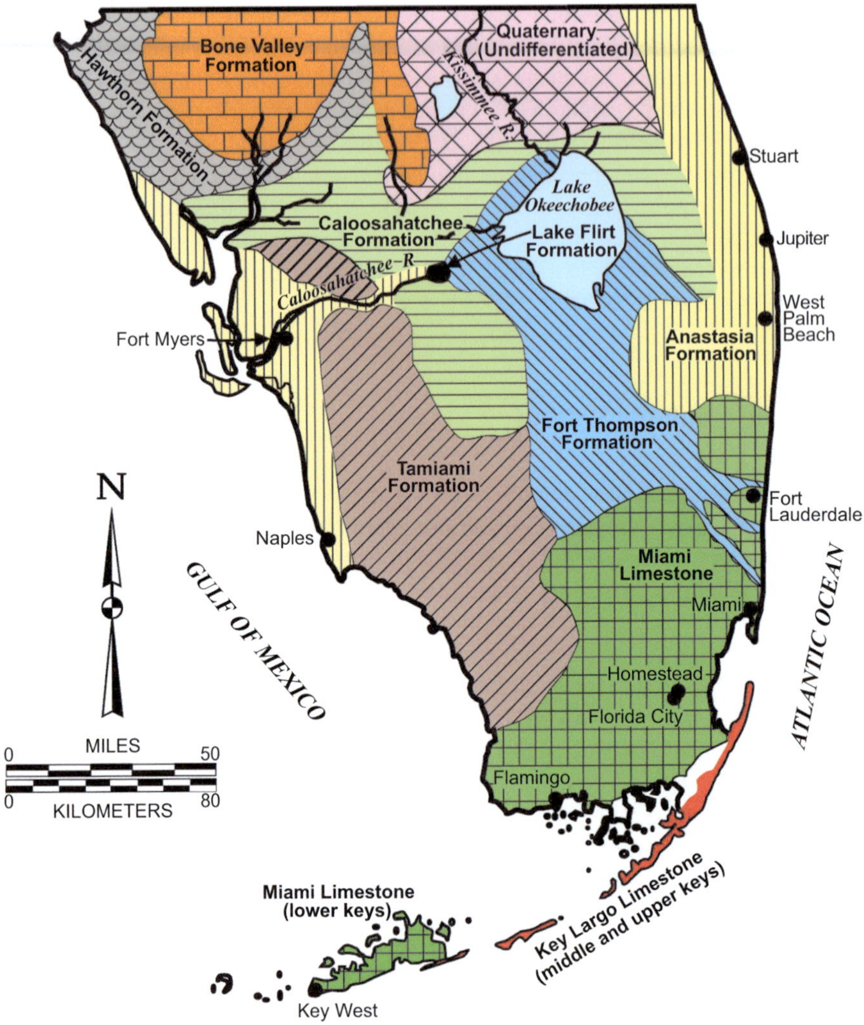

Fig. 1.3 A generalized map of south Florida's surface geologic exposures. (Source: Lodge 2017 after Gleason and Stone 1994, with permission of Taylor & Francis Group, LLC)

the Fort Thompson is overlain by Anastasia sands, so that the floor of the Everglades is not always rock. In the southern Everglades, the Fort Thompson is overlain by a western extension of the Miami Limestone, which contains abundant bryozoan fossils of shallow marine origin instead of the oolites of the same formation on the ACR. Beneath the central and southern Everglades, the Miami Limestone and underlying Fort Thompson formation are highly transmissive and comprise the Biscayne aquifer, which conducts groundwater from the southern half of the Everglades generally eastward, making the abundant groundwater supply used from Miami south to

Homestead and vicinity. The Biscayne aquifer also carries flows into Biscayne Bay, now greatly diminished due to area-wide drainage (Parker et al. 1955).

Older deposits surround the north end and the entire western edge of the Everglades. At the north end is the Caloosahatchee Formation. It is composed of highly variable materials but mainly sand and shells, and supports pine flatwoods and wetlands, depending on elevation. It is older than the Fort Thompson formation, evidenced by extending substantially beneath it. South of the Caloosahatchee Formation on the west side of the Everglades is the Tamiami Formation, deposited in Pliocene time. It is composed principally of limestone, but is highly variable, with sandstone as well as sandy and shelly layers, the uppermost being a discontinuous sandy layer sometimes designated as the Pinecrest member of the Tamiami Formation (Petuch and Roberts 2007). The Tamiami Formation is commonly exposed in the Big Cypress Swamp, often as hard cap rock. Its undulating surface accounts for much of the variability in vegetative cover in the Big Cypress, mostly wetlands having varying densities of cypress but also having significant cover by pinelands and upland hammocks. The Tamiami Formation also extends eastward under much of the Everglades, beneath the younger Miami Limestone and Fort Thompson formations, and is far thicker than the latter strata (Duever et al. 1986).

It is on this geologic setting that a veneer of very recent Everglades soils has been deposited (Gleason and Stone 1994).

1.4 Climate and Weather

In the Köppen classification, the climate of the Everglades and adjacent lands east and west is *tropical savanna* based on all months having an average daily temperature of 18 °C (about 64 °F) or warmer coupled with the distinct cooler dry season. From Lake Okeechobee and extending north through the remainder of Florida, the climate is *humid subtropical* based on cooler temperatures and a northward trend away from the pronounced dry season (Henry et al. 1994). The Everglades climate and proximity to the American tropics is also evidenced by many naturally occurring plants and animals of West Indian distribution that are not present farther north in Florida. The area's climate resembles the adjacent tropics more than it does neighboring locations in the continental United States. For the Everglades function as a wetland ecosystem, the alternating wet and dry seasonal cycle is the most important aspect of the climate. Precipitation is the main source of Everglades water. Unless otherwise cited, this summary is based on Duever et al. (1994).

Annualized Everglades rainfall data show both geographic and interannual variation. From 1951 to 1980, annual averages at five widely spaced stations ranged from 119 to 157 cm, representing geographic variability. A station at the northwest edge of the Everglades (Clewiston) was the lowest, the others being reasonably similar. Individual years averaged for all stations together, however, showed great variation, from 86 to 224 cm, representing droughts and wet years respectively.

Seasonal cycling of rainfall usually provides a summer-fall period of flooding that enables the growth of aquatic invertebrates and fishes, and winter-spring drying conditions that concentrate these organisms, thus enabling harvest by ecosystem predators, notably long-legged wading birds. The six-month dry season, November through April, accounts for only about 25% of the annual precipitation, with monthly averages in the range of 4–7 cm. Dry season precipitation is mostly associated with frontal boundaries, either southward-moving cold fronts or northward-moving warm fronts. Water levels recede slowly during the early and middle of the dry season but accelerate with warming and associated higher evapotranspiration toward the end. May is transitional, with wide fluctuations in rainfall due to variability in the onset of the wet season or extended dry conditions. Lightning due to increasing atmospheric convection in April and May sometimes sparks soil fires that reduce peat levels, especially in very dry years.

The four-month wet season, June through September, accounts for about 60% of the annual total. Wet-season monthly averages range between 16 and 25 cm. The wet season also shows two peaks for most locations, the larger in June and another in September, with a lull between. Most wet-season rainfall is from thunderstorms, but great excursions result from tropical cyclones of the "hurricane season," namely depressions, tropical storms, and hurricanes. The hurricane season begins concurrently with the normal wet season but extends later, officially designated as June 1 to November 30. Although tropical cyclones may bring excessive rain, long-term records show that tropical cyclones provide less of the annual rainfall than day-to-day summer thunderstorms. Like May, October is transitional and highly variable in rainfall, but normally has high water levels continuing from the wet season.

Departures in the Everglades region's normal seasonal rainfall pattern include droughts and multi-year flooding. Influences include sea-surface temperature cycles of both the tropical Atlantic and eastern Pacific, the latter best known as the El Niño Southern Oscillation (ENSO). ENSO influences weather patterns in Florida and Georgia, as both El Niño and the opposite, La Niña phases, related to warmer and cooler eastern Pacific surface temperatures, respectively. El Niño episodes are associated with increased winter and early spring rainfall from frontal systems, lower temperatures, and reduced tropical cyclones of summer and fall. La Niña has opposite but less obvious influences in southern Florida, being associated with dryer conditions, higher temperatures, and more frequent hurricanes. El Niño or La Niña recur at two to seven-year intervals. Combined, they affect less than a third of years, two thirds being described as the normal or "neutral phase" weather pattern (Fernald and Purdum 1998; Winsberg 2003).

Also superimposed on the normal seasonal cycles are freezes, more frequently associated with El Niño. Minimum temperatures in the Everglades region correlate with the distribution of many plants and animals. Five degrees C or less is lethal to many tropical species and has been used to define the limits of the tropics. In the Everglades region, this temperature criterion restricts tropical classification for ecology to the Florida Keys, the remainder of south Florida being ecologically "subtropical" (Armentano et al. 2002). Episodic freezes occur on the mainland, with increasing frequency northward. However, Tomlinson (2001) shows that many West

Indian tropical plants, particularly upland trees, occur on mainland south Florida in addition to the keys, their distribution correlating with an average January daily minimum temperature isotherm of 54 °F (12.2 °C). This isotherm is U-shaped and extends from Cape Canaveral on the Atlantic coast and running southward close to the east coast, then crossing the state through the southern Everglades and continuing back northward close to Florida's west coast to Tampa. Using more recent temperature data, Lodge (2017) shows that the northern portion of the January 54 °F isotherm has moved substantially south on both coasts, indicating probable climate change as increased temperature extremes, which is in agreement with a cold-related southward movement in citrus agriculture. In addition to the freeze protection provided by proximity to the ocean, interior water bodies including Lake Okeechobee, and wetlands when flooded, are important in protecting tropical vegetation. Thus, cold-protection for tropical plants changes by location with differences in vicinity flooding from year to year.

1.5 Everglades Plant Communities and Soils

A generalized cross-section that depicts the plant communities, soils, and underlying bedrock in the Everglades is shown in Fig. 1.4. It uses terminology specific to the Everglades. Unless otherwise credited, the descriptions below are based on Gunderson (1994), with references to predrainage conditions, water depths, and

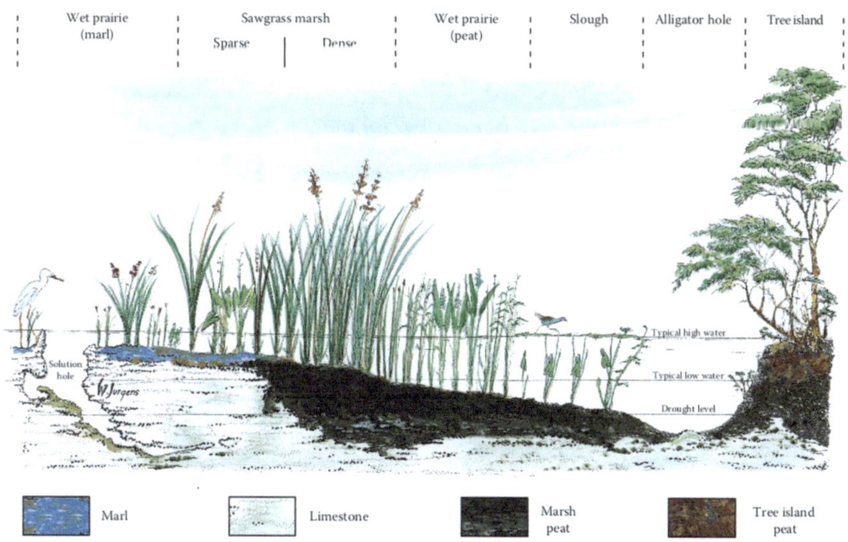

Fig. 1.4 A generalized, greatly compressed cross section of typical Everglades plant communities, water levels, and underlying soils and bedrock. (Source: Lodge 2017, with permission of Taylor & Francis Group, LLC)

hydroperiods based on McVoy et al. (2011). Common and scientific names of plants follow current usage (Wunderlin et al., 2016) thus may differ from names in the sources cited.

1. Sawgrass marshes. This most widespread Everglades community is overwhelmingly dominated by Jamaica swamp sawgrass (*Cladium jamaicense*), a sedge usually simply called "sawgrass." This species occurs where annual flooding (hydroperiod) is from 1 month to continuous over a few years (Olmsted and Loope 1984). It does not tolerate prolonged high water nor rising water soon after a fire because air tubes in dead leaves become flooded, eliminating oxygen for root respiration (Alexander 1971). Soil fires are also lethal, but sawgrass is highly tolerant of surface fire, recovering to full stature in 2 years. The species has very low phosphorus requirements; thus its success in the oligotrophic Everglades.

   Sawgrass marshes are described as either dense or sparse sawgrass, with few areas being intermediate. Dense sawgrass occurs on peat soils more than 1 m in depth, which characterized much of the predrainage peatland Everglades, including the Sawgrass Plains and ridges of the Ridge and Slough landscape (see Fig. 1.6). In those locations, the peat was of sawgrass origin and called *Everglades peat* (Gleason and Stone 1994). In predrainage times, dense sawgrass had an estimated average hydroperiod of 9–10 months with annual depths averaging −0.5 ft. at low water and 1.5 ft. at high, and the sawgrass attained a height to 3 m. Dense sawgrass has little other vegetation (Craighead 1971).

   Sparse sawgrass is less dense and occurs on marl or as a successional stage in degrading sloughs. It attains heights in the range of 0.8–1.5 m and may harbor numerous other marsh species such as gulf coast spikerush (*Eleocharis cellulosa*), lemon bacopa (*Bacopa caroliniana*), and marsh mermaidweed (*Proserpinaca palustris*). In the predrainage Everglades, sparse sawgrass likely occurred in the peat and marl transverse glades, and peripheral areas of the southern Everglades.
2. Sloughs. The word "slough" is used in Everglades literature for the plant community having open water and lily pads, primarily American white waterlily (*Nymphaea odorata*) but also spatterdock (*Nuphar advena*) and big floatingheart (*Nymphoides aquatica*). Except where degrading due to insufficient flow or reduced depth, sloughs lack emergent vegetation. "Slough" is also used in the names for geographic features in the Everglades and Big Cypress Swamp. An example is the southern Everglades "Shark River Slough" (see Fig. 1.6), a huge flow-way where the slough community is abundant together with sawgrass ridges and tree islands. It carries water toward its namesake, the Shark River, in the mangrove swamps. Use of the term "slough" in the Big Cypress Swamp is more varied than in the Everglades, designating deeper flow-ways that may be dominated by marsh or swamp forest, the latter alternatively called "strands." Examples of such named flow-ways in the Big Cypress are Okaloacoochee Slough, Mullet Slough, and Sweetwater Strand.

In the predrainage Everglades, sloughs were normally flooded continuously with annual depths averaging 1 ft. at low water and 3 ft. at high. Sloughs are underlain by a peat soil called *Loxahatchee peat*, primarily composed of the remains of American white waterlily (Dreschel et al. 2018; Gleason and Stone 1994). Recent advances in understanding the origin of ridge and slough landscape, including the abrupt edges between sawgrass ridges and sloughs, has shown that periodic water flow in the range of at least 2.5–5 cm/s is required. Such flows function to clear flocculent sediments from sloughs, preventing their accumulation, which would facilitate sawgrass encroachment. Sediments thus swept from sloughs would settle in sawgrass ridges where flows are slowed. The lack of sufficient flow has caused the loss of sloughs with the encroachment of sawgrass and other marsh vegetation in many areas of the Everglades (Larsen and Harvey 2010; Larsen and Harvey 2011; Harvey et al. 2017).

3. Wet Prairies (Peat). This low-stature, low-density graminoid marsh community is often characterized by gulf coast spikerush, maidencane (*Panicum hemitomon*), and Tracy's beaksedge (*Rhynchospora tracyi*), but numerous other species occur. These areas are thought to represent successional degradation of the slough community. Water depth and hydroperiod are intermediate between slough and sawgrass of the peatland Everglades. This community was common in dryer portions of the water conservation areas in the 1950s (Loveless 1959) as is the case today, apparently the result of lower water levels due to drainage (McVoy et al. 2011). This assumption is supported by the lack of peat representing this community in soil cores from numerous Everglades locations (Gleason and Stone 1994).
4. Wet Prairies (Marl). Wet prairies on marl soils occur principally in the southern Everglades on both sides of the Shark River Slough and along Taylor Slough (see Fig. 1.6). A commonly used name for this community, adopted here, is *marl prairie* (Olmsted and Loope 1984). Other names, applied to Everglades locations, include "southern Everglades marsh prairie" and "southern coast marsh prairie" (Davis 1943, Fig. 1.5), and "southeastern saline Everglades" (Egler 1952). In some locations, limestone solution holes and/or protruding pinnacle rock are common. Varying densities of short, scattered cypress may be present, mostly pond cypress (*Taxodium ascendens*), often called "hatrack cypress" (widely present in the Big Cypress Swamp) (Gunderson 1994). Wet prairie elevations are higher than peat-forming marshes and occur where the limestone bedrock is close to the surface. The hydroperiod is the shortest of all Everglades marshes. In predrainage times, the hydroperiod is estimated at 8–9 months, but more recent observations give a shorter range of 3–7 months (Gunderson 1994). Marl prairies have the highest species richness of the Everglades marshes, with over 100 species known to occur (Olmsted and Loope 1984), but are normally dominated by just two: sawgrass and hairawn muhly (*Muhlenbergia capillaris*). Most species are grasses and sedges, and community stature is less 1 m. Fire is frequent, restricting encroachment by shrubs, except in depressions, which may support tree islands, described below.

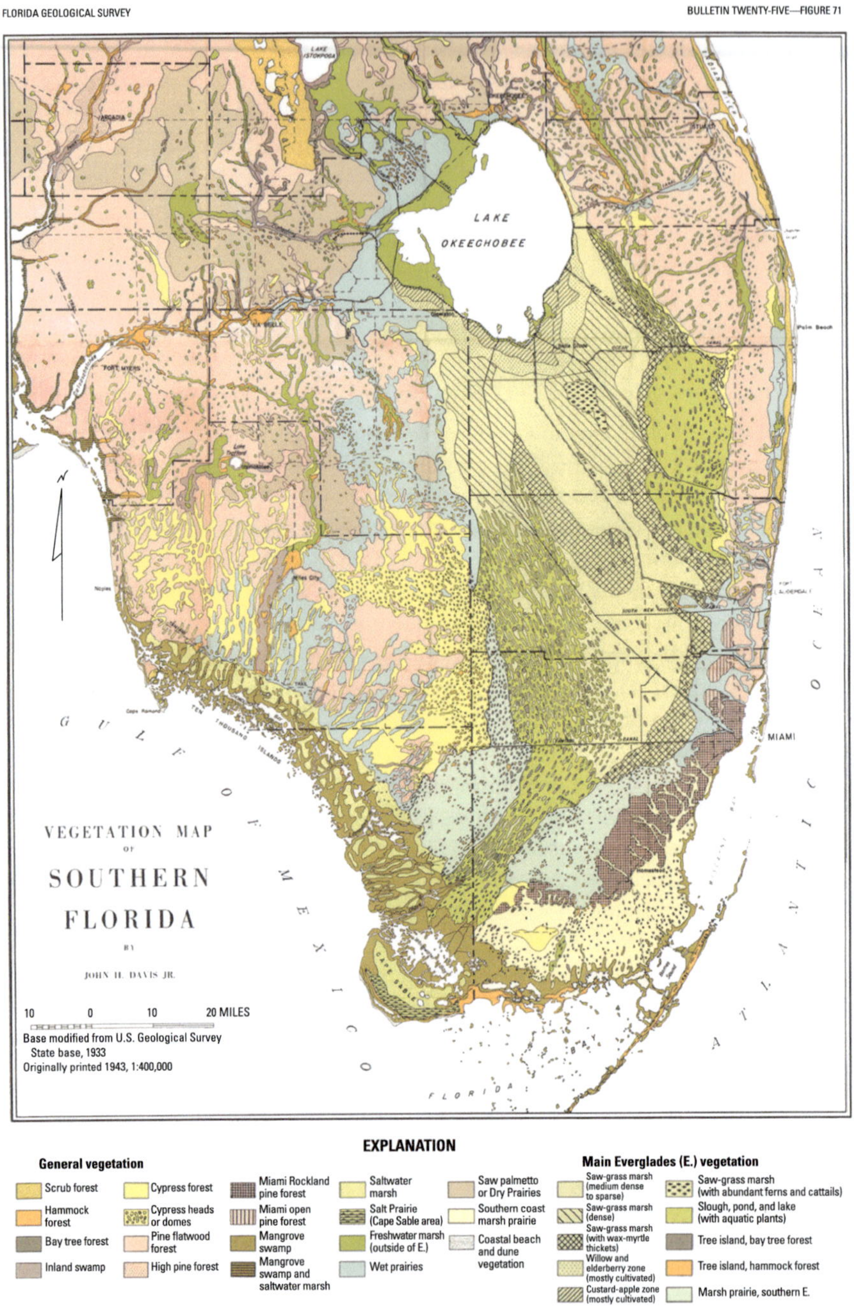

Fig. 1.5 Vegetation map of southern Florida by Davis (1943). (Source: provided courtesy of U.S. Geological Survey by Paul A. Conrads)

5. Periphyton. Periphyton is a complex community with over 700 microalgal and other taxa having been identified in the Everglades. Many are cyanobacteria (blue-green algae). Periphyton occurs in marsh communities as an envelope around plant stems or as an algal mat on the bottom, and it dominates primary production that is the base of the Everglades food web in marsh habitats (Gaiser 2009). Where dissolved calcium carbonate is abundant, periphyton also produces marl (Browder et al. 1994). Periphyton is highly efficient in sequestering and recycling phosphorus in freshwater areas where the concentration of phosphorus is 10 ppb or less but collapses rapidly as an effective community at significantly higher levels (Gaiser et al. 2005).
6. Ponds. Natural open water areas in the Everglades are the result of disturbances, primarily due to fire and/or alligator activity, and are temporary unless maintained. Alligators perpetuate such ponds, thus their common name "alligator hole." Alligator activities uproot vegetation and pile soil around pond perimeters. This activity also maintains the open waters of tidal creeks at the marsh-mangrove transition in the extreme southern Everglades. Vegetation responses include the establishment of Carolina willows around edges. Alligator holes (and creeks) are important in dry periods as feeding oases for wildlife such as wading birds and as refugia for surviving aquatic life available to repopulate marshes upon reflooding. Limestone solution holes in some areas also function as refugia (DeAngelis et al. 1997; Craighead 1971). Alligator holes are sites of concentrated nutrients, importantly phosphorus (Lodge 2017), promoting the frequent occurrence of cattails (*Typha* spp.) around them (Palmer and Mazzotti 2004). Cattails are unusual in unenriched areas of the Everglades (Gleason and Stone 1994).
7. Tree Islands. Tree islands in the Everglades are forested areas that appear as islands in contrast to the surrounding marsh. They are important to many ecological functions, including wildlife and nutrient dynamics, and some have been influenced by human occupation dating back to the birth of the ecosystem (Carr 2002). In addition to studies published in periodical literature, an entire book is dedicated to Everglades tree islands (Sklar and van der Valk 2002).

   Six kinds of tree islands occur in ridge-and-slough and marl prairies. Except for the heads of fixed tree islands, all are dominated by trees of temperate, North American origin. *Gandy* peat, the remains of woody vegetation, underlies tree islands that have been established long enough to accumulate such a layer. In turn, Gandy peat may overlie other substrates, described for particular kinds of tree islands, below. Most tree islands are much younger than the Everglades, their starts correlating with dry intervals in the climate record. In the case of certain large (fixed) tree islands, initiation to maturity took on the order of a thousand years. (Dreschel et al. 2018; Gleason and Stone 1994; Wetzel et al. 2017; Willard et al. 2002; Willard et al. 2006).

Most tree islands are loosely called "bayheads" in vernacular use because of the common presence of bay trees, namely red bay (Persea sp.)[2] and sweetbay (*Magnolia virginiana*), but several ecologically related species such as wax myrtle (*Myrica cerifera*) and dahoon (*Ilex cassine*) are also common. Tree islands evidencing these species are properly divided into three types: a. bayheads, b. strand tree islands, and c. fixed tree islands. Tree islands that are never called bayheads are cypress heads and willow heads, also described below. These descriptions are based on Gleason and Stone (1994) unless otherwise referenced:

(a) Bayheads in peatland Everglades. The tree islands that are universally called "bayheads" are small, typically ranging from 15 to 30 m across. Their perimeter vegetation is often a "skirt" of sawgrass but may also have shrubs or small trees including cocoplum (*Chrysobalanus icaco*) and pond apple (*Annona glabra*), contrasting with the interior bayhead species listed above. While many approximate a circle, they are commonly irregular, lacking any consistent pattern or orientation. In the peatland Everglades, they originate on patches of dislodged slough-bottom soil. Buoyancy results from gas accumulations in water lily rhizomes and roots, sometimes causing an area of soil to break free as a "battery" (also "sudd" or "pop-up"). The battery floats until lodged over other soil or against another tree island, the latter accounting for irregularly shaped bayheads. The battery provides a surface for germination and growth of wetland trees, shrubs, and herbs, and eventually a canopy is formed by trees that can endure flooding for a month or more. Bayheads are seldom anchored on bedrock thus are sometimes moved by strong winds. Eventually they develop a Gandy peat base over the parent Loxahatchee peat, thus often have three layers of different peats: Gandy peat over Loxahatchee peat, over the Everglades peat (where the battery initially lodged on a sawgrass ridge) (Brandt et al. 2002).

(b) Strand tree islands. Compared to bayheads, these tree islands are huge, up to 1600 m long, and elongated in the direction of historic water flow. They occur only over deep peat in the northern Everglades area that is now in the Loxahatchee National Wildlife Refuge (WCA-1). They have formed by succession on sawgrass ridges where protected from fire by adjacent sloughs for long periods and thus are underlain by Gandy peat formed over Everglades peat, unlike bayheads. Their elevation is only marginally higher than sawgrass ridges, lower than of bayheads. The canopy is dominated by dahoon (*Ilex cassine)*, which tolerates longer flooding than most bayhead species, and there is no significant difference in vegetative character over their length (Brandt et al. 2002).

[2]The name "red bay" is entrenched in Evergades literature as a bayhead species, referenced to the scientific name *Persea borbonia*. However, two varieties occur in the region and are now usually recognized as separate species: swamp bay (*P. palustris*), which tolerates shallow flooding for a month or more, is common in the Everglades; and red bay (*P. borbonia*), which is less tolerant of flooding and unusual or absent in the Everglades (Wunderlin et al. 2016; Nelson 2011). In lumping both together, a wide hydrologic tolerance is shown, as in Wetzel et al. 2017)

(c) "Fixed" tree islands. These are large, tear-drop shaped tree islands that have an elevated mineral base, a "head," at the upstream end based on the direction of predrainage water flow. Few occur over deep peat soils of the northern Everglades, but are common (except for recent demise, see Modern Everglades) in the central and southern Everglades. Trees and shrubs growing on the head, which is at least 1 m above the surrounding marsh, are typically upland species of the tropical hammock association, discussed below. They are intolerant of significant flooding in contrast to the suite of species of bayheads and strand tree islands. The mineral base of fixed tree island heads is small compared to the downstream expanse of the island, which grades from upland trees on the head to bayhead species on the near tail to the far tail of sawgrass, ferns, and shrubs, such as common buttonbush (*Cephalanthus occidentalis*), all tolerant of several months flooding (Willard et al. 2002).

All types of tree islands have soils with significantly higher phosphorus levels than the surrounding marsh, but fixed tree islands have extraordinary levels, related to processes on the head. Formerly, the mineral base of fixed tree island heads was assumed to be a prominence of bedrock. However, recent work has demonstrated that the head's larger, upland hammock trees greatly concentrate minerals from the underlying bedrock. The minerals then precipitate during dry seasons, forming a hard carbonate lens that differs from the Everglades bedrock. The concentration of phosphorus in fixed tree-island heads has been shown to be far greater, sometimes by more than 100 times, than found in the adjacent marsh. Progressively lower phosphorus levels, still significantly higher than in the adjacent marsh, are found through the tree-island near and far tails, consistent with downstream transport and supporting theory that fixed tree islands have grown from the initial head as a nucleus (Dreschel et al. 2018; Wetzel et al. 2011; Wetzel et al. 2017; Troxler et al. 2014). Further, numerous fixed tree island heads have midden material of Native American occupation, suggesting that the occurrence and distribution of fixed tree islands may have been significantly influenced by humans (Ardren et al. 2016; Graf et al. 2008).

(d) Marl-prairie bayheads. In contrast to bayheads of the peatland Everglades, bayheads in the marl prairies of the southern Everglades have not evolved from batteries and have somewhat different canopy flora, frequently having pond cypress (*Taxodium ascendens*) for example. They occur in depressions, typically as linear drainage pathways in the direction of historic flow but sometimes as individual pockets. Gandy peat fills these depressions to elevations 10–60 cm above the adjacent marsh, but with a lower, wetter moat-like margin that protects such bayheads from surface fires. These bayheads may normally lie directly over the limestone bedrock without intervening marl (Armentano et al. 2002; Gunderson 1994).

(e) Cypress heads. Also called cypress domes (when their profile fits that appearance), these tree islands occur in marl prairies but not in Everglades peatlands. They are dominated by pond cypress (Gunderson 1994). Like bayheads in marl prairies, they occupy depressions filled with peat but have an elevation lower than the marl prairie, thus with a longer hydroperiod. Other trees and shrubs that that tolerate the longer hydroperiods may occur, such as pond apple and water ash (*Fraxinus caroliniana*). Cypress heads and domes are abundant in the Big Cypress Swamp where they occur in rock depressions that are commonly circular and have open-water ponds in the center. Floristically similar linear corridors in the Big Cypress include cypress strands and sloughs (Ewel 1990; McVoy et al. 2011).

(f) Willow heads. Dominated by Carolina willow (*Salix caroliniana*), willow heads are small tree islands that usually occur where there has been a localized severe soil disturbance, as from peat fire, alligator activity, mechanized alterations, or changes in hydrology. They occur in both peatland and marl Everglades. Large-scale disturbances may result in a willow forest. Such occurrences of willow are not long-lasting landscape features, eventually succeeding to other plant communities (Armentano et al. 2002).

8. Plant communities surrounding the Everglades. Around the Everglades boundary, as well as limited areas within, are several widespread South Florida plant communities. Prominent among them are the following:

(a) Pinelands (based on Abrahamson and Hartnett 1990, unless otherwise referenced). The most widespread plant community of southern Florida is pineland, usually called pine flatwoods or just flatwoods, shown pink on the Davis vegetation map, Fig. 1.5. South Florida flatwoods are dominated by slash pine (*Pinus elliottii*), the only species reaching the canopy and only pine species present in the immediate vicinity of the Everglades. The slash pine tolerates hydric soils but not prolonged flooding. Where annual shallow flooding or soil saturation is brief, the understory is dominated by saw palmetto (*Serenoa repens*), an indicator of *mesic* pine flatwoods. *Hydric* pine flatwoods are flooded for a month or two and lack saw palmetto. Their elevation is only marginally higher than bordering graminoid marshes, such as the wet prairies north of the Big Cypress Swamp, coded blue on the Davis map, Fig. 1.5.

On the Miami Rock Ridge in Miami-Dade County, flatwoods occur on the widely exposed Miami oolite limestone and are called "pine rocklands." They are indicated as cross-hatched maroon on the Davis map (Fig. 1.5). The pine rocklands extend into Everglades National Park, the area known as Long Pine Key, which lies west of Taylor Slough. This community's flora, generally of temperate North American origin, is highly diverse and noted for large numbers of rare and endemic taxa (Gunderson 1994). All pine flatwoods communities are a fire-climax. Fire inhibits invading hardwoods in the process of succession. Without fire, flatwoods succeed to the hardwood hammock community. The natural fire return interval is 5–10 years (Gunderson 1994).

(b) Tropical hardwood forests. Tropical hardwood forests usually exist as localized patches called "tropical hardwood hammocks," the climax community of the Everglades region where elevations are above seasonal flooding. Concurrent with a warming climate, an influx of tropical vegetation, mostly trees, started in southern Florida about 6000 years ago, originating from the West Indian-Caribbean region (Long 1984). In the Florida Keys and the central and southern Everglades, nearly 90% of the species are of tropical affinity (Tomlinson 2001). Most tropical hammock trees and shrubs have seeds that are bird-dispersed (Olmsted and Loope 1984). More temperate species and fewer tropical species are found in the hammocks of the northern Everglades and its vicinity, but with tropical species reaching farther north close to the Atlantic and Gulf coasts (see Climate and Weather).

(c) Mangrove swamps and salt marshes. Surface waters in the southern Everglades mostly flow via the Shark River Slough to the Gulf of Mexico but also to western Florida Bay. Lesser flows move through the much smaller Taylor Slough to central Florida Bay. Historically, substantial surface and groundwater also flowed into Biscayne Bay, now reduced and routed through canals (Parker et al. 1955). Where the freshwater flows reach tidal influences of brackish salinities, mangrove swamp forests dominate. They are narrow along Biscayne Bay due to abrupt shoreline elevations but are expansive over the flatter topography from the north shoreline of central Florida Bay northwest to Naples, Florida. The Big Cypress Swamp provides fresh water to the western portion. Salt marshes, mostly dominated by black rush (*Juncus roemerianus*), also occur but cover less area, unlike their expanses in temperate climates. These communities are shown on the Davis vegetation map, Fig. 1.5, as medium brown (mangrove swamps) and yellow-green (salt marshes).

In the Everglades region, mangrove swamps as well as mangrove islands in Florida Bay and the Ten Thousand Islands, are populated by three unrelated intertidal species: the red mangrove (*Rhizophora mangle*), distinctive with its arching, adventitious "prop" roots; the black mangrove (*Avicennia germinans*) with its dark bark and abundant, finger-like pneumatophores rising from its roots to above tide levels; and the white mangrove (*Laguncularia racemosa*) with its light bark and light green oval leaves. A fourth species, buttonwood (*Conocarpus erectus*), in the same plant family as the white mangrove, is associated but occurs from higher intertidal elevations into uplands. All these species are of tropical origin and occur as forests only around the southern half of the Florida peninsula. They are best developed along tidal creeks and rivers that receive flows from the Shark River Slough. Maximum mangrove heights are 100 ft. (30 m) but normally less than half that. Farther north, mangroves exist as scattered, cold-pruned individuals (Davis 1940; Odum and McIvor 1990).

The mangrove swamps of southern Florida are renowned for productivity that nurtures the coastal food web including many game and food fishes and shellfishes, most notably the pink shrimp (*Farfantepenaeus duorarum*). In the past, mangroves along tidal creeks in the transition from the Shark River Slough to mangrove forests were the location of the huge wading-bird rookeries of the Everglades. Mangrove islands are also important for roosting and nesting by numerous birds including roseate spoonbills and brown pelicans (USFWS 1999; Odum and McIvor 1990).

1.6 Modifications of the Everglades

The history of modifications to the Everglades from its predrainage configuration shown on Fig. 1.6, is documented by Godfrey and Catton (2011), Grunwald (2006), Light and Dineen (1994), McCally (1999), McVoy et al. (2011), and summarized by Lodge (2017). The modifications fall into four groups:

1. Earliest Changes. Relatively minor reductions in the level of Lake Okeechobee were made in 1883 by connecting it to the Caloosahatchee. This work was done mainly for the purpose of draining lands in the upper Kissimmee River system, not for draining the Everglades where it had little effect.
2. Everglades Drainage District. Major projects by Florida's Everglades Drainage District, created in 1907, were made for agricultural development of the northern Everglades. Works included an earthen dike along part of Lake Okeechobee and excavation of four drainage canals from the lake through the Everglades to carry water to the southeast coast. They were the West Palm Beach, Hillsboro, North New River, and Miami canals, initially completed by 1917. Completed in 1926, the St. Lucie Canal was connected to the St. Lucie River at Stuart, Florida, making the shortest outlet from the lake. These works greatly reduced water levels in the Everglades but were insufficient for hurricanes in 1926 and 1928, which caused extensive flooding and loss of life associated with breaches of the lake's dike.
3. Federal Intervention. With intervention by the federal government in 1930, a substantial dike was constructed around nearly all of Lake Okeechobee and upgrades of the connecting canals were made, including their outlets from the lake and associated navigation features. The dike was completed in 1937 and named the Herbert Hoover Dike. In 1947, it was successfully tested by two hurricanes, but flooding caused by canal releases was extensive. The drainage system could not accommodate excessive rainfall, and in dry periods over-drained the Everglades, causing muck fires.
4. C&SF Project. Congress authorized the Central and Southern Florida Project for Flood Control and Other Purposes (C&SF Project) in 1948. It became the massive modification of the entire system that exists today, including the following components: (1) conversion of much of the northern Everglades into the Everglades Agricultural Area (EAA), (2) creation of three "water conservation areas" covering the remaining northern and most of the central Everglades, (3) severing and drainage of large areas of the central and southern Everglades west of Fort Lauderdale and Miami for land development, and (4) most importantly, extensive increases in capacity for stormwater conveyance from the interior, including the Kissimmee-Lake Okeechobee watershed, to the coast. The project's principal role was a flood-control system. The water conservation areas became relatively flat impoundments, with interior flows reduced compared to the predrainage condition, superimposed on the original sloping landscape. Then newly created, Everglades National Park (1947) became the downstream, recipient of an artificially controlled system of canals, levees, and pumps. Many features of the C&SF Project together with the predrainage Everglades boundary

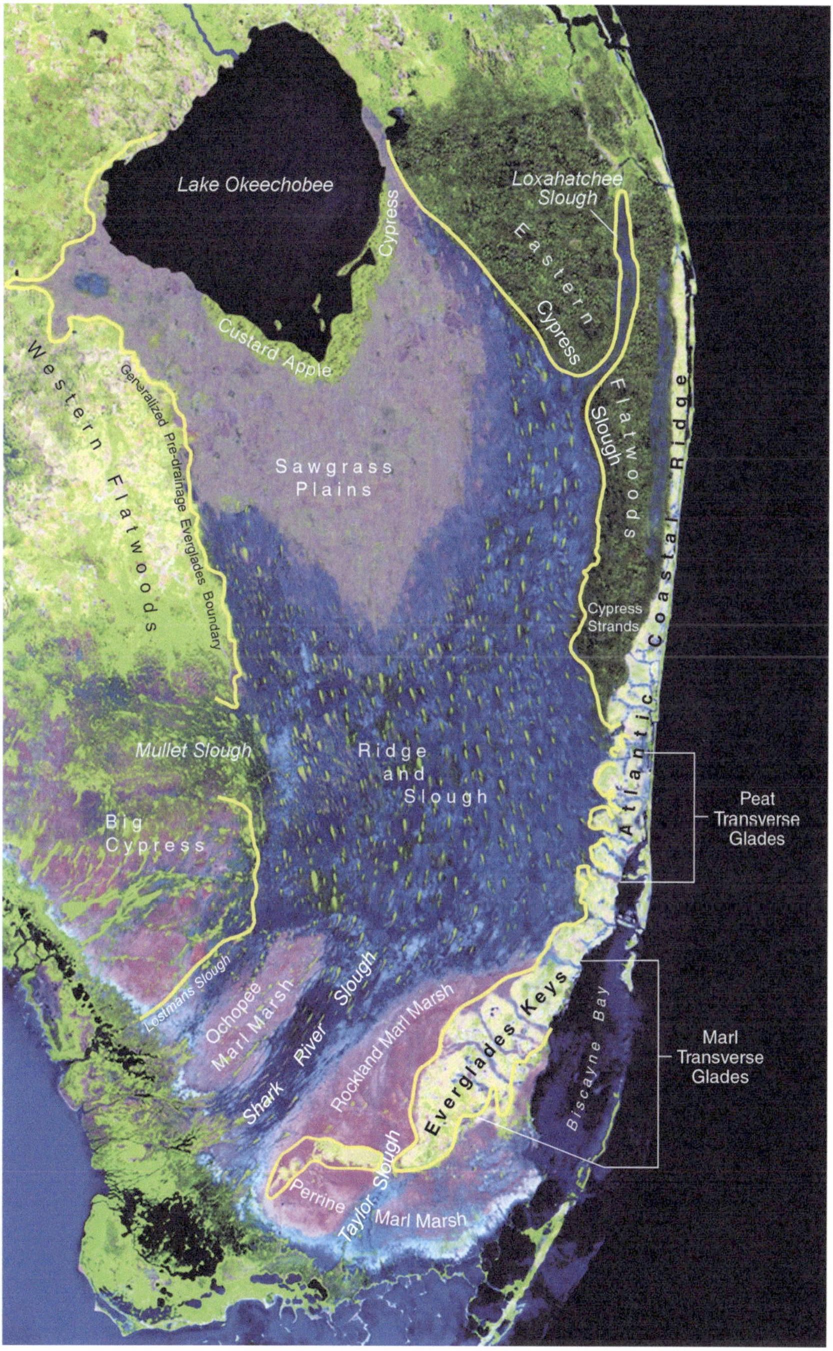

Fig. 1.6 Simulated satellite image of the predrainage Everglades. The yellow line is the border of the Everglades detailed in the text. (Source: McVoy et al. 2011, courtesy of the South Florida Water Management District)

are shown on Fig. 1.7. The altered flow patterns can be seen in comparing Figs. 1.8 and 1.9. As an overall summary, the area of the predrainage Everglades was 1,106,000 ha and is now 623,200 ha, or 56% of the original (McVoy et al. 2011). It is the C&SF Project components that are in the process of being modified under the ongoing Comprehensive Everglades Restoration Plan (CERP) (USACE and SFWMD 1999), discussed at the end of this chapter.

1.7 Predrainage Everglades Landscapes: The Basis for Understanding Change

The Everglades plant communities, described above, occurred in defined landscapes in the predrainage ecosystem. These landscapes and their current condition are shown on Figs. 1.6 and 1.7, respectively. Figures 1.8 and 1.9 show the respective changes in the patterns of water flow. The predrainage landscapes and their percent cover of the Everglades are described below based on McVoy et al. (2011) except where alternately referenced, followed by an explanation of differences between the predrainage condition of McVoy et al. (2011) and the vegetation mapping of Davis (1943) (see Fig. 1.5):

1. Custard Apple and Cypress Swamp Forests. These communities bordered the southern and southeastern shore of Lake Okeechobee and covered only about 1% of the Everglades. They were dominated by pond apple, regionally called "custard apple," but it graded northeastward into cypress. The pond apple swamp forest was 2–3 miles wide but narrowed and disappeared westward so that sawgrass of the Sawgrass Plains bordered the lake's southwest shore. Short rivers, called "blind rivers," led into this area from Lake Okeechobee, but divided and ended in wetlands so that early explorers could not find passages deeper into the Everglades. Where pond apple and cypress were dominant, this landscape was flooded nearly the entire year. It was underlain by Okeechobee muck and more limited Okeelanta peaty muck to the south (see qualification at the end of this section regarding the "willow and elderberry zone"). Okeechobee muck was a layered sequence derived from the woody remains of the trees, aquatic vegetation, and sediments from Lake Okeechobee, with Everglades peat below. This swamp forest no longer exists, entirely developed for agriculture and urban uses of the Everglades Agricultural Area (EAA) (Dreschel et al. 2018; Gleason and Stone 1994).
2. Sawgrass Plains. This landscape lay generally south of the lake and swamp forest described above. It covered about 24% of the Everglades. A small area on the west side of the lake was the headwaters of the Caloosahatchee while the remainder conducted water southward (see flow pattern on Fig. 1.8). The Sawgrass Plains area was overwhelmingly dominated by dense sawgrass. It lacked open-water rendering it nearly impenetrable for early explorers. It was underlain by Everglades peat, as deep as 14 ft near Lake Okeechobee, deposited over lime-

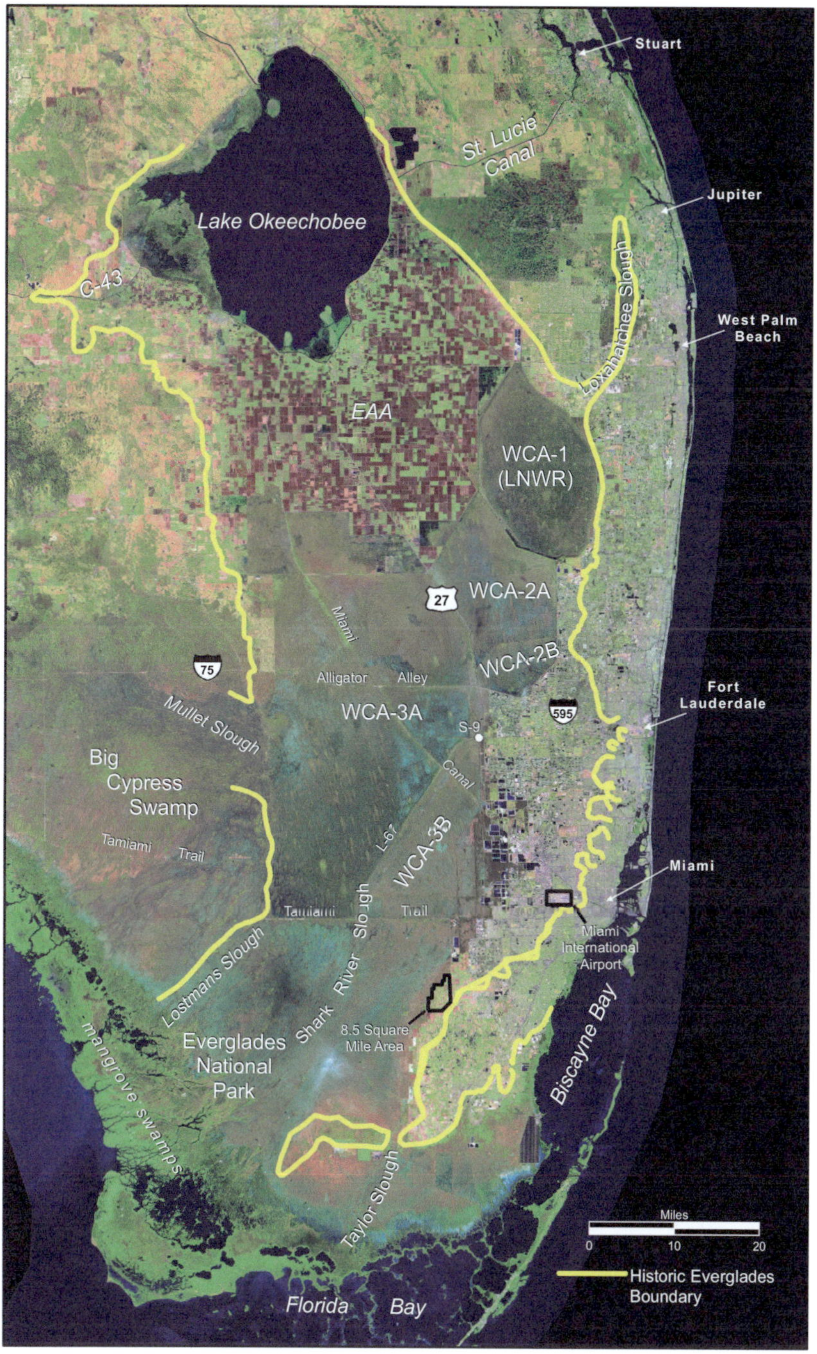

Fig. 1.7 South Florida 2004 satellite image showing the predrainage Everglades boundary, water conservation areas (WCAs), the Everglades Agricultural Area (EAA), canal connections from Lake Okeechobee to the St. Lucie River and to the Caloosahatchee (C-43), and other features. (Courtesy of the South Florida Water Management District with modifications by Lodge 2017, with permission of Taylor & Francis Group, LLC)

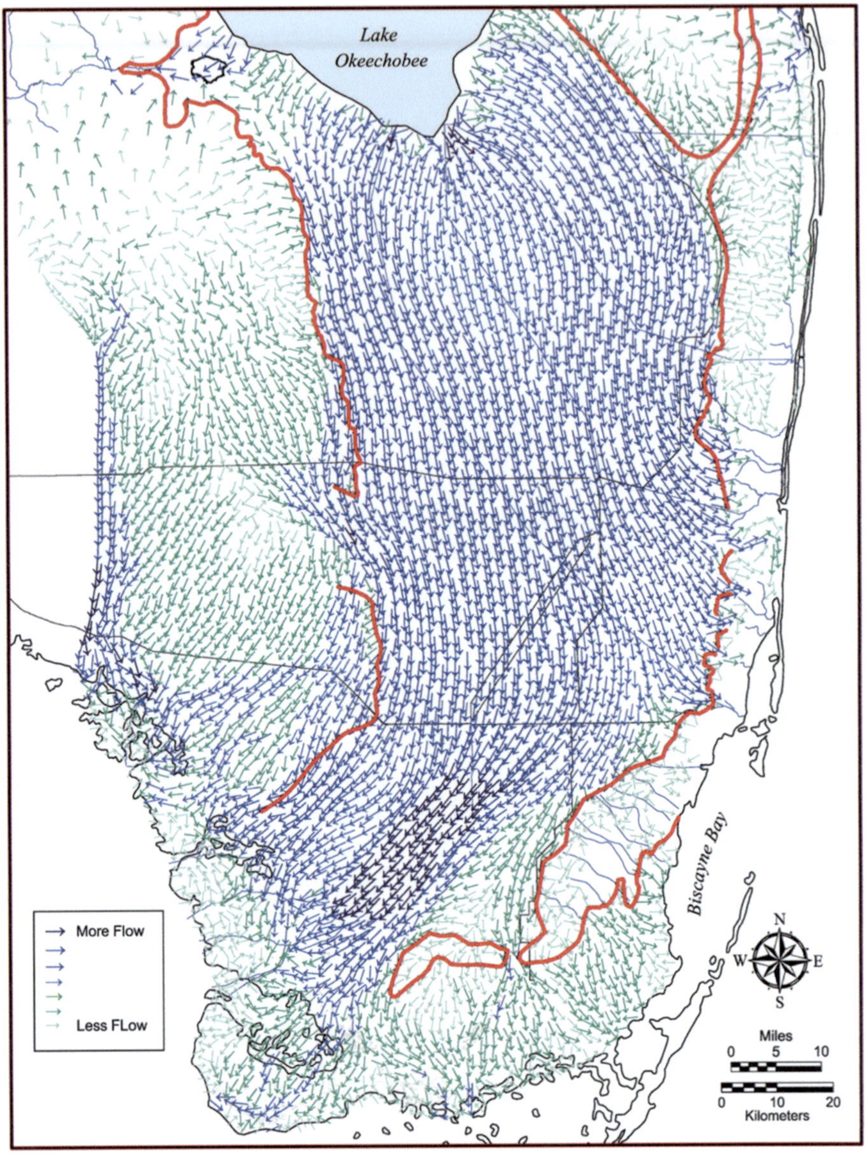

Fig. 1.8 Surface flows of the predrainage Everglades generated by the Natural System Regional Simulation Model v3.3. Arrow colors/sizes depict flow magnitude. Selected modern roads and levees are shown for orientation. The red line is the border of the Everglades. (Source: Courtesy of the South Florida Water Management District)

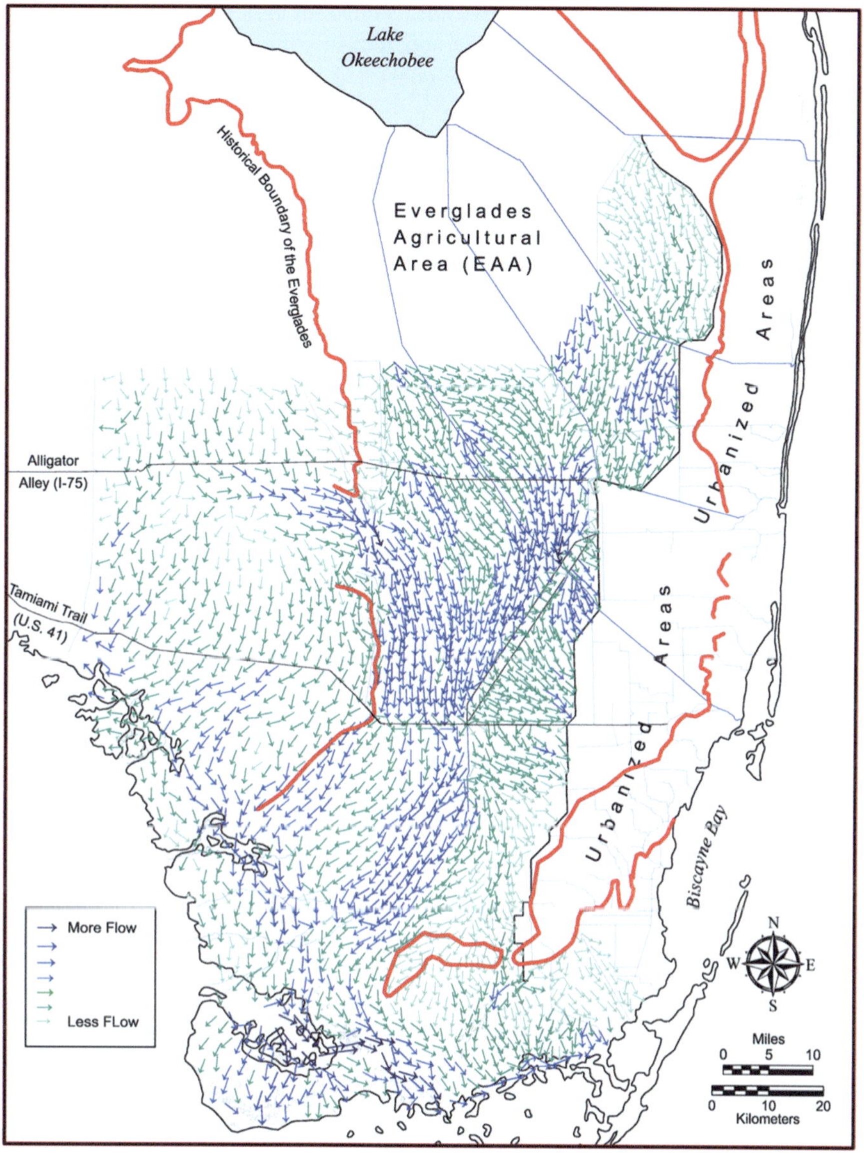

Fig. 1.9 Surface flows of the current Everglades, 2010, generated by the Glades-LECSA Model. Arrow colors/sizes depict flow magnitude. The red line is the predrainage Everglades border. (Source: Courtesy of the South Florida Water Management District)

stone bedrock. Its drained soils have been valued for agriculture, primarily sugar cane, but have subsided greatly. As apparent in comparing Figs. 1.6 and 1.7, essentially all of this landscape was converted into the EAA (Dreschel et al. 2018; Gleason and Stone 1994).

3. Ridge and Slough. This was the most extensive landscape, covering about 55% of the Everglades. It extended from the edges of the sawgrass plains through the southern Everglades where it narrowed from its maximum width of about 40 miles to the six-mile-wide Shark River Slough (see Fig. 1.6). Flows in the Shark River Slough, now in Everglades National Park, fed into tidal creeks and rivers of the coastal mangrove swamps. The ridge-and-slough landscape was characterized by elongated sawgrass ridges of dense sawgrass, open-water sloughs, and tree islands, all arranged in obvious alignment with the general north-to-south water flow direction. The sawgrass ridges were about 2 ft higher than the sloughs. In the predrainage Everglades, the area covered by sloughs and sawgrass ridges were similar, and the Loxahatchee peat of sloughs was not desirable for agriculture thus the limitation of the EAA to the sawgrass plains. Where tree islands were most abundant in the northeastern Everglades, they occupied about 14% of ridge and slough landscape but less in areas farther south, and the numbers of tree islands as well as their aerial cover has greatly decreased since predrainage times (Sklar and van der Valk 2002).
4. Marl Marshes of the Southern Everglades. These totaled about 19% of the Everglades. Tree islands (discussed above) occurred in depressions. Regional names were Ochopee Marl Marsh, the Rockland Marl Marsh, and the Perrine Marl Marsh (see Fig. 1.6).
5. Transverse Glades. These pathways of overland flow were in low areas of the Atlantic Coastal Ridge and totaled about 2% of the Everglades. The northern ones drained directly from the ridge-and-slough landscape and had peat soils. The southern ones had marl soils and were continuations from the southern Everglades marl marshes, cutting across the south end of the Atlantic Coastal Ridge before flowing to Biscayne Bay or into coastal marl marshes (Parker et al. 1955).
6. Changes in understanding predrainage landscapes. For many decades after its publication, the southern Florida vegetation map of Davis (1943) was the standard of reference (see Fig. 1.5). It was the base used in most Everglades research works including Davis and Ogden (1994). More recent work by McVoy et al. (2011) has demonstrated that Davis had mapped important changes resulting from regional drainage, then in place for more than 20 years. Notable differences between the interpretations of McVoy et al. (2011) and Davis (1943) are the following:

   (a) Ridge-and-slough landscape. Davis identified two separate areas of the Everglades having the above described characteristics of this landscape: a northern area in what is now in Water Conservation Area-1 (WCA-1) and much of WCA-2A, and a southern area now mostly in WCA-3 and Everglades National Park. Various sawgrass marshes lay between them. McVoy et al.

(2011) conclude that these two areas were formerly united, and that water levels between them had been sufficiently reduced by the early drainage canals (completed in 1917) to allow sawgrass to encroach (succession) into former sloughs, obscuring those original landform features and leading Davis to define various categories of sawgrass-dominated cover.

(b) The "willow and elderberry zone" of Davis, located south of his "custard-apple zone." Dominated by these low trees and shrubs in 1940, it was on soil described as "Okeelanta peaty muck," a profile sequence that was mostly Everglades peat (Gleason and Stone 1994). With this and a lack of earlier recognition, McVoy et al. (2011) believe that this community did not exist in the predrainage Everglades but was another result of early drainage and attendant succession of vegetation.

The Davis map (Fig. 1.5) continues to be of value in understanding the magnitude of early changes in the Everglades and because much of south Florida's vegetation – including the southern Everglades and areas outside of the Everglades – is credibly mapped for predrainage time.

1.8 Wildlife and Food Web

It was the biological values of the Everglades that led to the creation of Everglades National Park (Beard 1938; Fairchild 1945). Of those values it was the abundance of wading birds – herons, egrets, ibises, the wood stork, and the roseate spoonbill – that attracted the most attention. Together with the American alligator, this suite of wading birds represents the top predators and keystone species of the Everglades (Davis and Ogden 1994). The Florida panther (*Puma concolor coryi*) is a regional apex predator, but more important in lands surrounding the Everglades, notably in the Big Cypress Swamp, where its primary prey, white-tailed deer (*Odocoileus virginianus*) and feral pigs (*Sus scrofa*), are more abundant, not in the marsh-dominated Everglades (Smith and Bass 1994).

The seasonal rainfall cycle with consequent flooding and drying over large areas of marshes is the most important driver of the Everglades food web, which is mainly based on primary production by periphyton (Gaiser 2009). Flooding promotes growth of a prey base, mainly small forage fishes and aquatic invertebrates. Alternatively, fishes and mobile invertebrates follow receding water as dry seasons progress, forming a concentrated prey base accessible to wading birds. Variables that influence the available biomass include 1) access of fishes and invertebrates to re-flooded marshes from refugia such as alligator holes; 2) the duration of flooding, with longer periods producing more biomass, and 3) the sequence of drying for prey concentration. For the latter, reversals in water levels due to unusual dry-season rain events and/or water management cause dispersal of prey, diminishing availability to predators (DeAngelis et al. 1997). Accordingly, there is high variability in wading-bird success in different years and different locations (Frederick and Ogden 2001;

Ogden 1994). Overall, Everglades wading-breeding populations have fallen greatly, concurrent with breeding areas having relocated north from the historic rookeries at the mangrove-marsh transition area to the impounded water conservation areas (Ogden 1994), but with some notable recent improvements for a few species (Cook 2016; reviewed by Lodge 2017).

1.9 Threatened and Endangered Species

Sixty-eight federally listed species (threatened or endangered) and many more state-listed species occur within the greater Everglades ecosystem (or South Florida ecosystem). Of the federally listed species, at least 17 were shown to be potentially affected by Everglades restoration (USFWS 1999; USACE and SFWMD 1999). Since those publications, a few species have been added and the listing status has changed for some. Most of these listed species occur in habitats surrounding the Everglades, including uplands and coastal estuarine and marine areas, not specifically in the freshwater marshes and tree islands of the Everglades. Examples include the Florida manatee (*Trichechus manatus latirostris*, endangered) and the American crocodile (*Crocodylus acutus*, threatened), both primarily living in coastal estuarine and marine habitats but occasionally entering canals passing through the Everglades; the Florida panther (endangered) addressed above; and the tiny polygala (*Polygala smallii*, endangered) one of several listed plants that occur in uplands, this species found in pine rocklands and other upland habitats of southeast Florida.

In the freshwater wetlands of the Everglades, three federally listed species have drawn considerable attention due to potential impacts of Everglades restoration and conflicting needs between the individual species: the Everglade snail kite (*Rostrhamus sociabilis plumbeus*, endangered), the wood stork (*Mycteria americana*, threatened), and the Cape Sable seaside sparrow (*Ammodramus maritimus mirabilis*, endangered) (Brosnan 2007; Orians et al. 1992; Pimm et al. 2002; USACE 2014; USFWS 1999).

1.10 Invasive Species

Many species have invaded the Everglades as a result of recent human activities. Invasive plants have included the native southern cattail and many non-indigenous (exotic) species. South Florida now has 829 naturalized, non-indigenous plant taxa of which 166 have been designated as invasive –7% of the total flora (Lodge 2017). While extensive and well documented damage has been caused by species such as Brazilian pepper (*Schinus terebinthifolia*), Australian pine (*Casuarina equisetifolia*), and waterthyme (*Hydrilla verticillata*), relatively few non-indigenous species have become problems specifically in the marshes. In addition to invasive plants, numerous exotic animal species have impacted the Everglades (Lodge 2017). Selected problematic species are briefly described here.

Southern cattail (*Typha domingensis*): Cattails were not common in the predrainage Everglades (McVoy et al. 2011; Gleason and Stone 1994), but the southern cattail has spread into the ecosystem, initially in limited areas during early drainage (Davis 1943); but following completion of the Everglades Agricultural Area in 1959 (Light and Dineen 1994), phosphorus-enriched stormwater discharges to the ecosystem caused extensive infestations in the water conservation areas. The infestations precipitated a 1988 lawsuit by the federal government against the State of Florida over violation of the state's water quality standards, namely causing an "… imbalance of aquatic flora and fauna…." (Lodge 2017). Numerous subsequent studies have detailed how elevated phosphorus facilitates cattail invasion of Everglades communities, creating dense infestations that have poor water quality and limited ecosystem value (Davis 1994; Noe and Childers 2007).

Melaleuca (*Melaleuca quinquenervia*): This highly fire-resistant tree from Australia was introduced in 1906. By the 1970s, it was widely invading short-hydroperiod wetlands such as sawgrass and cypress throughout South Florida including the littoral zone of Lake Okeechobee. Unlike many invasive exotics, it can overtake undisturbed natural plant communities, advancing by wind-blown seeds (Schmitz et al. 1997). Melaleuca has formed dense stands of tall trees that damage wildlife use, such as by wading birds (O'Hare and Dalrymple 1997). Costly herbicidal control has been widely used. Biological-control insects have been introduced with some success, notably in reducing tree stature (SFWMD 2016).

Old World climbing ferns (*Lygodium* spp.): The first introduced in 1958, two species (*L. japonicum*, and *L. microphyllum*) have become invasive in South Florida, including tree islands of the Everglades. Both are vines that cover herbaceous plants and shrubs as well as advancing to the canopy of tall trees such as cypress. Deleterious impacts result from their dense cover and formation of "fire ladders" that carry surface fires into forest canopies. Tree islands in northern Everglades have been heavily impacted, requiring expensive but only marginally effective control (Brandt and Black 2001; Schmitz et al. 1997). Infestations have moved south and west to the central and southern Everglades and Big Cypress Swamp. Two modestly effective biological controls have been released and others are under development (SFWMD 2016).

Burmese python (*Python molurus*): Individuals of this large constrictor from Southeast Asia were occasionally encountered in South Florida from the 1980s to 2000. Rapid population growth then began, with annual numbers removed from Everglades National Park increasing from a few prior to 2002, to nearly 400 in 2009. A small and brief reduction in python numbers occurred in 2010, credited to a January freeze that killed many large individuals. Severe declines of mammal populations (marsh rabbit, raccoon, opossum, bobcat, etc.) have correlated with the increasing python population, and those species in addition to white-tailed deer and alligators have been documented in python diets in the park (Dorcas et al. 2012). Birds taken have included the wood stork (Dove et al. 2011). No credible python population estimate exists for South Florida, and the secretive habits of the Burmese python have made control efforts ineffective (Lodge 2017).

1.11 Everglades Restoration

Initiatives to improve the Everglades have taken many paths. Early attempts to improve conditions in Everglades National Park involved schedule changes in the operation of water release structures, with mixed success. Deterioration of wetland conditions in the water conservation areas has been attributed mostly to phosphorus enrichment, but more recently the lack of sufficient flow to maintain remaining ridge-and-slough landscape. Reducing phosphorus loads from the Everglades Agricultural Area stemmed from legal action (see Cattail section under Everglades Invasive Species above) that resulted in establishing a 10 ppb standard for phosphorus in the "Everglades Protection Area" (the water conservation areas and Everglades National Park), and consequential construction of stormwater treatment areas (STAs) under the State of Florida's Everglades Construction Project. To date, six numbered STAs have been constructed in the EAA with a total effective treatment area of 57,000 acres. Phosphorus levels have been greatly reduced but full compliance has not been achieved (Lodge 2017). More recently, sulfur enrichment coming principally from agriculture has gained attention. It is associated with elevated soil sulfide levels and elevated methylmercury (extensively treated in this volume), the latter biomagnified in Everglades fish and other wildlife to levels of concern for wildlife survival and human health (Scheidt and Kalla 2007). There has been no program for sulfur reduction (Lodge 2017).

The largest initiative is the joint federal-state Comprehensive Ecosystem Restoration Plan (CERP), authorized by Congress in 1999. It involves many aspects, the main approach being an overhaul of the South Florida water-control system (principally the C&SF Project features) to ensure the right quality, timing, and distribution of water. Its implementation is beyond the scope of this overview. However, the literature and internet abound with relevant information. Of high importance are biannual reviews by the Committee on Independent Scientific Review of Everglades Restoration Progress of the National Academies of Sciences (National Research Council 2016).

References

Abrahamson WG, Hartnett DC (1990) Pine flatwoods and dry prairies. In: Myers RL, Ewel JJ (eds) Ecosystems of Florida. University of Central Florida Press, Orlando

Alexander TR (1971) Sawgrass biology related to the future of the Everglades ecosystem. Soil Crop Sci Soc Florida Proc 31:72–74

Ardren T, Lowry JP, Memory M, Flanagan K, Busot A (2016) Prehistoric human impact on tree island lifecycles in the Florida Everglades. The Holocene 26(5):772–780

Armentano TV, Jones DT, Ross MS, Gamble BW (2002) Vegetation pattern and process in tree islands of the southern Everglades and adjacent areas. In: Sklar FH, van der Valk A (eds) Tree islands of the Everglades. Kluwer Academic Publishers, Boston

Beard DB (1938) Wildlife reconnaissance, Everglades National Park project. U.S. Department of the Interior, National Park Service, Washington, DC

Brandt LA, Black DW (2001) Impacts of the introduced fern, *Lygodium microphyllum*, on the native vegetation of tree islands in the Arthur R. Marshall Loxahatchee National Wildlife Refuge. Fla Sci 64(3):191–196

Brandt LA, Silveira JE, Kitchens WM (2002) Tree islands of the Arthur R. Marshall Loxahatchee National Wildlife Refuge. In: Sklar FH, van der Valk A (eds) Tree islands of the Everglades. Kluwer Academic Publishers, Boston

Brosnan D (project lead) (2007) Everglades multi-species avian ecology and restoration review: final report. Sustainable Ecosystems Institute, Portland

Browder JA, Gleason PJ, Swift DR (1994) Periphyton in the Everglades: spatial variation, environmental correlates, and ecological implications. In: Davis SM, Ogden JC (eds) Everglades: the ecosystem and its restoration. St. Lucie Press, Delray Beach

Carr RS (2002) The archaeology of Everglades tree slands. In: Sklar FH, van der Valk A (eds) Tree islands of the Everglades. Kluwer Academic Publishers, Boston

Cook MI (ed) (2016) South Florida wading bird report, vol 21. South Florida Water Management District, Everglades Systems Assessment, W. Palm Beach. This publication is part of a series of annual reports

Craighead FC Sr (1971) The natural environments and their succession, The Trees of Southern Florida, vol 1. University of Miami Press, Coral Gables

Davis JH Jr (1940) The ecology and geologic role of mangroves in Florida. Carnegie Institution, publication no. 517, Washington, DC

Davis JH Jr (1943) The natural features of southern Florida, especially the vegetation, and the Everglades. The Florida Geological Survey, Geological bulletin no. 25, Tallahassee

Davis SM, Ogden JC (eds) (1994) Everglades: the ecosystem and its restoration. St. Lucie Press, Delray Beach

DeAngelis DL, Loftus WF, Trexler JC, Ulanowicz RE (1997) Modeling fish dynamics and effects of stress in a hydrologically pulsed ecosystem. J Aquat Ecosyst Stress Recover 6:1–13

Dorcas ME, Willson JD, Reed RN, Snow RW, Rochford MR, Miller MA, Meshaka WE Jr, Andreadis PT, Mazzotti FJ, Romagosa CM, Hart KM (2012) Severe mammal declined coincide with proliferation of invasive Burmese python in Everglades National Park. PNAS 108(7):2418–2422

Douglas MS (1947) The Everglades: river of grass. Rinehart, New York. (60th anniversary ed. with "Afterword" by M. Grunwald, 2007, 2010. Pineapple Press, Inc., Sarasota)

Dove CJ, Snow RW, Rochford MR, Mazzotti FJ (2011) Birds consumed by the invasive Burmese python (Python molurus bivittatus) in Everglades National Park, Florida, SA. Wilson J Ornithol 123(1):126–131

Dreschel TW, Hohner SM, Aich S, McVoy CW (2018) Peat soils of the Everglades of Florida, USA. In: Topcuoğlu B, Turan M (eds) Peat. ISBN 978-953-51-5805-9, In-press

Duever MJ, Carlson JE, Meeder JF, Duever LC, Gunderson LH, Riopelle LA, Alexander TR, Myers RL, Spangler DP (1986) The big cypress national preserve. National audubon society research report no. 8. National Audubon Society, New York

Duever MJ, Meeder JF, Meeder LC, McCollom JM (1994) The climate of South Florida and its role in shaping the Everglades ecosystem. In: Davis SM, Ogden JC (eds) Everglades: the ecosystem and its restoration. St. Lucie Press, Delray Beach

Egler FE (1952) Southeast saline Everglades vegetation in Florida, and its management. Veg Acta Bot III(Fasc. 4-5):213–265

Ewel KC (1990) Swamps. In: Myers RL, Ewel JJ (eds) Ecosystems of Florida. University of Central Florida Press, Orlando

Fairchild D (1945) Are the Everglades worth saving? National Parks Magazine, Jan.-Mar. 1945 (reprinted from an address by Dr. Fairchild at the 1929 annual meeting of the American Forestry Association)

Fernald EA, Purdum ED (1998) Water resources Atlas of Florida. Institute of Science and Public Affairs, and Florida State University, Tallahassee

Fletcher R, Robertson E, Reichert B, Cattau C, Wilcox CR, Zweig C, Jeffery B, Olbert J, Pias K, Kitchens W (2015) Snail kite demography 5-year report, prepared for the U.S. Army Corps of Engineers, Environmental Branch, Jacksonville, FL. by U.S. Geological Survey, FL Cooperative Fish and Wildlife Research Unit and Department of Wildlife Ecology and Conservation, University of Florida, Gainesville

State of Florida (1845) A plat exhibiting the state of the Surveys in the State of Florida with references. Surveyor General's Office, St. Augustine

Frederick PC, Ogden JC (2001) Pulsed breeding of long-legged wading birds and the importance of infrequent severe drought conditions in the Florida Everglades. Wetlands 21(4):484–491

Gaiser E (2009) Periphyton as an indicator of restoration in the Florida Everglades. Ecol Indic 9S:37–45

Gaiser EE, Trexler JC, Richards JH, Childers DL, Lee D, Edwards AL, Scinto LJ, Jayachandran K, Noe GB, Jones RD (2005) Cascading ecological effects of low-level phosphorus enrichment in the Florida Everglades. J Environ Qual 34:717–723

Gawlik DE, Gronemeyer P, Powell RA (2002) Habitat-use patterns of avian seed dispersers in the central Everglades. In: Sklar FH, van der Valk A (eds) Tree islands of the Everglades. Kluwer Academic Publishers, Boston

Glaser PH, Hansen BCS, Donovan JJ, Givnish TJ, Strickler CA, Volin UC (2013) Holocene dynamics of the Florida Everglades with respect to climate, dustfall, and tropical storms. PNAS 110(43):17211–17216

Gleason PJ, Brooks HK, Cohen AD, Goodrick R, Smith WG, Spackman W, Stone P (1984) The environmental significance of Holocene sediments from the Everglades and saline tidal plain. In: Gleason PJ (ed) Environments of South Florida: present and past II. Miami Geological Society, Miami

Gleason PJ, Stone P (1994) Age, origin, and landscape evolution of the Everglades peatland. In: Davis SM, Ogden JC (eds) Everglades: the ecosystem and its restoration. St. Lucie Press, Delray Beach

Godfrey MC, Catton T (2011) River of interests: water management in South Florida and the Everglades, 1948–2010. Historical Research Associates, Inc., under contract to the USACE (for sale by the Superintendent of Documents, U.S. Government Printing Office, Washington, DC)

Goodrick RL (1984) The wet prairies of the northern Everglades. In: Gleason PJ (ed) Environments of South Florida: present and past II. Miami Geological Society, Miami. (also in 1974 edition)

Graf MT, Schwadron M, Stone PA, Ross M, Chmura GL (2008) An enigmatic carbonate layer in Everglades tree island peats. Eos 89(12):117–124

Grunwald M (2006) The swamp: the Everglades, Florida, and the politics of paradise. Simon & Schuster, New York

Gunderson LH (1994) Vegetation of the Everglades: determinants of community composition. In: Davis SM, Ogden JC (eds) Everglades: the ecosystem and its restoration. St. Lucie Press, Delray Beach

Harshberger JW (1913) Phytogeographic map of South Florida. Hand drawn, 24 × 26 inches, 1:500,000. Wagner Free Institute of Science, Philadelphia

Harvey JW, Wetzel PR, Lodge TE, Engle VC, Ross MS (2017) Role of a naturally varying flow regime in Everglades vegetation. Restor Ecol 25:S27–S38. https://doi.org/10.1111/rec.12558

Heisler LD, Towles T, Brandt LA, Pace RT (2002) Tree island vegetation and water management in the central Everglades. In: Sklar FH, van der Valk A (eds) Tree islands of the Everglades. Kluwer Academic Publishers, Boston

Henry JA, Portier KM, Coyne J (1994) The climate and weather of Florida. Pineapple Press, Inc., Sarasota

Hogarth PJ (2015) The biology of mangroves and seagrasses, 3rd edn. Oxford University Press, Oxford

Ives LJC (1856) Military map of the Peninsula of Florida South of Tampa Bay. U.S. War Department, Washington, DC

Kirby RR, Hobbs CH, Mehta AJ (1989) Fine sediment regime of Lake Okeechobee, Florida. A study funded by the SFWMD as part of a Lake Okeechobee phosphorus dynamics study. Coastal and Oceanographic Engineering Department, University of Florida, Gainesville
Larsen LG, Harvey JW (2010) How vegetation and sediment transport feedbacks drive landscape change in the Everglades and wetlands worldwide. Am Nat 176(3):E66–E79
Larsen LG, Harvey JW (2011) Modeling of hydroecological feedbacks predicts distinct classes of landscape pattern, process, and restoration potential in shallow aquatic systems. Geomorphology 126:279–296
Light SS, Dineen JW (1994) Water control in the Everglades: a historical perspective. In: Davis SM, Ogden JC (eds) Everglades: the ecosystem and its restoration. St. Lucie Press, Delray Beach
Lodge TE (2017) The Everglades handbook: understanding the ecosystem, 4th edn. CRC Press, Boca Raton
Long RW (1984) Origin of the vascular flora of Southern Florida. In: Gleason PJ (ed) Environments of South Florida: present and past II. Miami Geological Society, Miami. (also in 1974 edition)
Loveless CM (1959) A study of the vegetation in the Florida Everglades. Ecology 40(1):1–9
McCally D (1999) The Everglades: an environmental history. University Presses of Florida, Gainesville
McPherson BF (1984) The big cypress swamp. In: Gleason PJ (ed) Environments of South Florida: present and past II. Miami Geological Society, Miami. (also in 1974 edition)
McVoy, C.W., Said, W.P., Obeysekera, J., Vanarman, J.A., Dreschel, T.W. (2011). Landscapes and hydrology of the predrainage Everglades. University Press of Florida, Gainesville
Missimer TM (1984) The geology of South Florida: a summary. In: Gleason PJ (ed) Environments of South Florida: present and past II. Miami Geological Society, Miami. (also in 1974 edition)
National Research Council (2016) Progress toward restoring the Everglades: the sixth biennial review – 2016. The National Academies Press, Washington, DC
Nelson G (2011) The trees of Florida: a reference and field guide, 2nd edn. Pineapple Press, Inc., Sarasota
Noe GB, Childers DL (2007) Phosphorus budgets in Everglades wetland ecosystems: the effects of hydrology and nutrient enrichment. Wetl Ecol Manag 15:189–205
Odum WE, McIvor CC (1990) Mangroves. In: Myers RL, Ewel JJ (eds) Ecosystems of Florida. University of Central Florida Press, Orlando
Ogden JC (1994) A comparison of wading bird nesting colony dynamics (1931–1946 and 1974–1989) as an indication of ecosystem conditions in the Southern Everglades. In: Davis SM, Ogden JC (eds) Everglades: the ecosystem and its restoration. St. Lucie Press, Delray Beach
O'Hare NK, Dalrymple GH (1997) Wildlife in southern Everglades wetlands invaded by melaleuca (Melaleuca quinquenervia). Bull Fla Mus Nat Hist 41(1):1–68
Olmsted I, Loope LL (1984) Plant communities of Everglades National Park. In: Gleason PJ (ed) Environments of South Florida: present and past II. Miami Geological Society, Miami
Orians GH, Bean M, Lande R, Loftin K, Pimm S, Turner RE, Weller M (1992) Report of the advisory panel on the Everglades and endangered species, Audubon conservation report no. 8. National Audubon Society, New York
Palmer M, Mazzotti FJ (2004) Structure of Everglades alligator holes. Wetlands 24(1):115–122
Parker G (1984) Hydrology of the pre-drainage system of the Everglades in southern Florida. In: Gleason PJ (ed) Environments of South Florida: present and past II. Miami Geological Society, Miami
Parker GG, Ferguson GE, Love SK, and others (1955) Water resources of southeastern Florida, with special reference to the geology and groundwater of the Miami area. U.S. Geological Survey Water-Supply Paper 1255, Washington, DC
Petuch EJ, Roberts CE (2007) The geology of the Everglades and adjacent areas. CRC Press, Boca Raton
Pimm SL, Lockwood JL, Jenkins CN, Curnutt JL, Nott MP, Powell RD, Bass OL Jr (2002) Sparrow in the grass: a report on the first ten years of research on the Cape Sable Seaside Sparrow

(Ammodramus maritimus mirabilis). Report under cooperative agreements CA5280-7-9016 (to the Univ. of Tennessee, Principal Investigator S.L. Pimm) and CA5280-0-0010 (to the University of California at Santa Cruz, Principal Investigator J.L. Lockwood)
Pomaea Project, Inc. (2013) Literature review of Florida apple snails and snail kites, and recommendations for their adaptive management. Final report. Submitted to National Park Service, Everglades National Park, by The Pomacea Project, Inc., Pensacola, Florida
Rader RB (1994) Macroinvertebrates of the northern Everglades: species composition and trophic structure. Fla Sci 57(1,2):22–33
Richardson CJ (2008) The Everglades experiments: lessons for ecosystem restoration. Springer, New York
Said WP, Brown MC (2013) Hydrologic simulation of the predrainage greater Everglades using the natural system regional simulation model v3.5.2. South Florida Water Management District, Water Resources Division, Water Supply Bureau, Hydrologic and Environmental Systems Modeling Section, West Palm Beach
Scheidt DJ, Kalla PI (2007) Everglades ecosystem assessment: water management and quality, eutrophication, mercury contamination, soils and habitat: monitoring for adaptive management: a R-EMAP status report. USEPA Region 4, Athens, GA. EPA 904-R-07-001
Schmitz DC, Simberloff D, Hofstetter RH, Haller W, Sutton D (1997) The ecological impact of nonindigenous plants. In: Simberloff D, Schmitz DC, Brown TC (eds) Strangers in paradise: impact and management of nonindigenous species in Florida. Island Press, Washington, DC
SFWMD (2008) South Florida environmental report, executive summary. South Florida Water Management District, W. Palm Beach, pp 24–25
SFWMD (2016) South Florida environmental report. South Florida Water Management District, W. Palm Beach
Sklar, F.H., van der Valk, A. (eds.) (2002). Tree islands of the Everglades. Kluwer Academic Publishers, Boston
Smith TR, Bass OL Jr (1994) Landscape, white-tailed deer, and the distribution of Florida panthers in the Everglades. In: Davis SM, Ogden JC (eds) Everglades: the ecosystem and its restoration. St. Lucie Press, Delray Beach
Steinman AD, Havens KE, Carrick HJ, VanZee R (2002) The past, present, and future hydrology and ecology of Lake Okeechobee and its watersheds. In: Porter JW, Porter KG (eds) The Everglades, Florida Bay, and Coral Reefs of the Florida Keys: an ecosystem source book. CRC Press, Boca Raton
Tomlinson PB (2001) The biology of trees native to tropical Florida, 2nd edn. Harvard Printing and Publications Services, Allston
Troxler TG, Coronado-Molina C, Rondeau DN, Krupa S, Newman S, Manna M, Rice RM, Sklar FH (2014) Interactions of local climatic, biotic and hydrogeochemical processes facilitate phosphorus dynamics along an Everglades forest-marsh gradient. Biogeosciences 11:899–914
USACE (2014) Final integrated project implementation report and environmental impact statement, Central Everglades Planning Project (CEPP). U.S. Army Corps of Engineers, Jacksonville District
USACE (2016) Western Everglades restoration project at: www.saj.usace.army.mil/Missions/Environmental/Ecosustem-Restoration/Western-Everglades-Restoration-Project/
USACE and SFWMD (1999) Central and Southern Florida project comprehensive review study final integrated feasibility report and programmatic environmental impact statement. Final report, appendices, and ATLSS
USFWS (1999) South Florida multi-species recovery plan. U.S. Fish and Wildlife Service, Southeast Region, Atlanta
USFWS (2012) Endangered and Threatened wildlife and plants; Reclassification of the continental U.S. breeding population of the Wood Stork from Endangered to Threatened. Fed Regist 77(247):75947–75966
U.S. Senate (1911) Everglades of Florida: acts, reports, and other papers, state and national, relating to the Everglades of the State of Florida and their reclamation. 62nd Congress, 1st

session, Senate document no. 89. Government Printing Office, Washington, DC. Notes: (1) This document was reprinted verbatim (undated, about 1998) as South Florida in Peril: a study in bureaucratic self-deception by Florida Classics Library, Port Salerno, Florida (ISBN 0-912451-43-2); and (2) This document contains the "Report of Buckingham Smith, ESQ, on his Reconnoissance[sic] of the Everglades (1848)" with appendix, pp 46–66

Vignoles C (1823) Observations upon the Floridas. Reproduction of the 1823 edition with prefatory material, introduction, and index added (1977 copyright). Bicentennial Floridiana Facsimile Series. The University Presses of Florida, Gainesville, 1977. A legible fold-out copy of Vignoles' 1823 map is included

Wanless HR, Parkinson RW, Tedesco LP (1994) Sea level control on stability of Everglades wetlands. In: Davis SM, Ogden JC (eds) Everglades: the ecosystem and its restoration. St. Lucie Press, Delray Beach

Wetzel PR, Sklar FH, Coronado-Molina CA, Troxler TG, Krupa SL, Sullivan PL, Ewe S, Price RM, Newman S, Orem WH (2011) Biogeochemical processes on tree islands in the greater Everglades: initiating a new paradigm. Crit Rev Environ Sci Technol 41(S1):670 701

Wetzel PR, Sah JP, Ross MS (2017) Tree islands: the bellwether of Everglades ecosystem function and restoration success. Restor Ecol 25:S71-75. https://doi.org/10.1111/rec.12428

Willard DA, Holmes CW, Korvela MS, Manson D, Murray JB, Orem WH, Towles DT (2002) Paleoecological insights on fixed tree island development in the Florida Everglades: I. environmental controls. In: Sklar FH, van der Valk A (eds) Tree islands of the Everglades. Kluwer Academic Publishers, Boston

Willard D, Bernhardt CE, Holmes CW, Landacre B, Marot M (2006) Response of Everglades tree islands to environmental change. Ecol Monogr 76(4):565–583

Willard D, Bernhardt CE (2011) Impacts of past climate and sea level change on Everglades wetlands: placing a century of anthropogenic change into a late-Holocene context. Clim Chang 107:59–80

Winsberg MD (2003) Florida weather, 2nd edn. University Press of Florida, Gainesville

Wunderlin RP, Hansen BF, Frank AR, Essig FB (2016) Atlas of Florida Vascular Plants (http://florida.plantatlas.usf.edu/). [S.M. Landry and K.N. Campbell (application development), USF Water Institute.] Institute for Systematic Botany, University of South Florida, Tampa

Chapter 2
The Evolution of the Everglades as a Perturbed Ecosystem

Darren G. Rumbold

Abstract Much has been written about the origin and evolution of the Everglades (for review, see Chap. 1, this volume). The objective here is not to restate what has already been written but instead to highlight the environmental features and processes, both natural and those altered by humans that combine to make the Everglades extraordinarily susceptible to mercury.

Keywords Altered hydrology · Agriculture · Soil amendments · Eutrophication · Biogeochemistry

2.1 Natural System

The historic Everglades extended over an area approximately 60–80 km wide by 145 km long, from the south shore of Lake Okeechobee to the mangrove estuaries of Florida Bay (Chap. 1, this volume). Yet, to understand the Everglades, one must also consider Lake Okeechobee and its source waters, e.g., Kissimmee River and several creeks (which have a combined drainage area of over 1 million ha; Jones 1948). This is sometimes termed the Kissimmee-Lake Okeechobee-Everglades (KLOE) Watershed (Chap. 1, this volume).

The Everglades has been described as a depression in the limestone comprising the surficial geology of the region that has filled with organic matter and sedimentary deposits (for review of processes creating basin, see Gleason and Stone 1994). The formation of the Everglades peatland from partly decayed plant materials began about 5000 years before present (YBP) following a change in climate leading to a more stable sea level (Gleason and Stone 1994). The various peat types and the specific plant materials they arose from is reviewed by Jones (1948) and summa-

D. G. Rumbold (✉)
Department of Marine and Environmental Sciences, Florida Gulf Coast University, Fort Myers, FL, USA
e-mail: drumbold@fgcu.edu

C. D. Pollman et al. (eds.), *Mercury and the Everglades. A Synthesis and Model for Complex Ecosystem Restoration*,
https://doi.org/10.1007/978-3-030-20070-1_2

rized in Chap. 1 (this volume). Not surprisingly, materials leached from this accumulating plant material also generated dissolved organic matter (DOM). Although highly variable (both in quantity and quality, as discussed later), surface waters of much of the KLOE watershed are naturally highly colored with organic matter (Parker et al. 1955; for review of DOMs role in the mercury (Hg) cycle, see Chap. 4, Volume II).

The Everglades has a generally subtropical climate (for review, see Chap. 1, this volume) with only occasional freezes associated with winter cold fronts (for review, see Duever et al. 1994). Tall convective thunderstorms (12–16 km) that extend into the upper troposphere are the major source of rainfall (for the influence these storms have on Hg atmospheric deposition, see Chap. 3, this volume) along with hurricanes and infrequent rain associated with winter frontal systems (average yearly rainfall ranges from 119 cm to 157 cm; Duever et al. 1994; the reader is referred to Chaps. 3 and 4, this volume for in-depth discussion of wet and dry Hg deposition). The wet season extends from May to October (75% of the rainfall) while the dry season occurs from November to April (25% of the rainfall). However, there is great spatial and temporal variability in rainfall with very wet years interspersed with droughts (Duever et al. 1994).

The Everglades is topographically flat with elevations generally less than 6 m (NGVD). The ground surface gently slopes from north to south with an average gradient of 2.8 cm/km (Parker et al. 1955). Consequently, the frequent intense, but short, rainfall events characteristic of the region result in extended periods of flooding that was augmented by drainage from Lake Okeechobee. When it filled from direct rainfall and inflow from the large watershed to the north, Lake Okeechobee historically overflowed along much of the southern rim into the Everglades because it had no well-defined outlets (Jones 1948). These waters would then sheet flow slowly southward to Florida Bay.

Davis (1943) produced the first comprehensive map of vegetation patterns (Chap. 1, this volume, Fig. 1.5) and reported that most of the original Everglades marsh was covered by sawgrass and other grasses (the reader is referred to Chap. 1, Volume II for in-depth discussion on susceptibility of marshes for Hg methylation). Gunderson (1994) describes the evolution of species endemic to the region and the floral assemblages of the Everglades. The unique vegetation communities are thought to have been selected because they flourished within an oligotrophic environment characteristic of extremely low levels of nutrients (Davis 1994). Historically, atmospheric precipitation had been the primary source of nitrogen (N) and phosphorus (P) to the Everglades (SFWMD 1992; Davis 1994). As discussed later, such ombrotrophic peatlands produced from plants grown under these conditions are generally nutrient poor (for review, see Koch and Reddy 1992 and references therein).

2.2 Anthropogenic Modifications to the System

The first major effort to drain the Everglades began in 1881 with a contract between Hamilton Disston and the Trustees of the Internal Improvement Fund (Jones 1948; for review, see Chap. 1, this volume). While Disston began connecting a series of lakes in the upper Kissimmee basin and dug a canal connecting the Caloosahatchee River and Lake Okeechobee, it was the creation of the Everglades Drainage District (EDD) in 1907 that produced major changes to the KLOE watershed (for review, see Chap. 1, this volume). By 1917, the EDD had completed construction of four major canals from Lake Okeechobee through the Everglades to Atlantic Ocean: the Miami, North New River, Hillsboro, and West Palm Beach Canals (Light and Dineen 1994; Chap. 1, this volume). By 1931, 440 miles of canals had been dug, 47 miles of constructed levees and 16 lock and dams built (Jones 1948). However, a dry spell between 1931 and 1945 revealed that this system engineered to drain the Everglades worked *too* well (Finkl and Makowski 2017; for ranks of droughts in Miami, Kissimmee and Florida as a whole, see Parker et al. 1955). The 1938–39 drought allowed saltwater intrusion into a Miami municipal well field, which prompted intensive investigations of water resources in southeastern Florida (Jones 1948). Water consumption for communities along the east coast, from West Palm Beach to Key West, was already at 50 million gallons per day (mgd) at that time (Parker et al. 1955; for a review of how the population went from 50,000 in 1845, when Florida gained statehood, to what it is today, 21 million, see Solecki et al. 1999). This over-drainage also resulted in frequent peat fires which prompted the creation of the Everglades Fire Control District in 1939. While the initial focus had been on draining the Everglades, Resource managers began to realize that "conservation and control of water was important for preserving the organic soils, irrigation of the farm crops, and for replenishment of subsurface storage from which the municipal supplies are pumped" (Jones 1948). For anyone that remained unconvinced that further water management was necessary, the great flood of 1947 demonstrated that besides over draining the system during dry periods, the system could not accommodate excessive rainfall (Finkl and Makowski 2017). Consequently, in 1948 and 1949 both the federal and state governments, respectively, took further steps to manage water. This led to creation of the Central and Southern Flood Control District (later to become the South Florida Water Management District) by the state to serve as the local sponsor to work with the US. Army Corps of Engineers. They immediately established the Central and Southern Florida Project for Flood Control and other Purposes (CSF Project) which began construction of perimeter levees and flood control structures that would define both the Water Conservation Areas (WCAs, encompassing approximately a million acres or 3500 km^2) and the Everglades Agricultural Area (EAA, which today encompasses approximately 2872 km^2; Redfield et al. 1999).

Although native Americans had been farming in south Florida for thousands of years, farming by white pioneers began only in the early 1900s along the shores of Lake Okeechobee on what is sometimes called Torry muck (also known as

Okeechobee muck, i.e., one of three general categories of highly organic mucks found in the region that are differentiated by mineral content; Snyder and Davidson 1994). Later, farming expanded further south into areas of sawgrass muck (i.e., Loxahatchee peat) but encountered difficulties (Snyder and Davidson 1994). Many of these difficulties had been predicted in a 1915 report by the U.S. Department of Agriculture that questioned the value of the soils (Jones 1948). Later it was shown that to grow anything other than sawgrass on the sawgrass muck would require "heavy application of phosphate and potash, and light application of certain elements such as copper, manganese, and zinc, usually in the sulfate form" (Jones 1948, pg 72; for a review, see McCally 1999). Reports of experiments at the Everglades Agricultural Experimental Station in the late 1920s, for example, showed that the addition of 50–100 pounds of manganese sulfate added to the commercial fertilizer (per acre) would make farming more profitable (Allison 1928 as cited by McCally 1999; also see Jones 1948). Farmers are reported to have quickly embraced these recommendations for truck crops and sugarcane (Jones 1948; McCally 1999). Later, it was recommended that agricultural sulfur (comprised of 98% elemental sulfur) be added as a soil amendment to reduce soil pH and enhance uptake of phosphorus and solubilize micronutrients (at a rate of 500 lbs to 2 tons per acre; Bottcher and Izuno 1994; for review of current sugarcane fertilizer recommendations, see Morgan et al. 2015; for review of sulfur sources and relation to the Hg cycle, see Chaps. 2 and 3, Volume II). Large acreage farms of sugarcane began appearing in the 1920s (125-acre farm in 1920, Snyder and Davidson 1994). Although plagued with numerous obstacles, including quota limitations for sugarcane production and the 1928 hurricane, agricultural production in the region advanced during the 1930s and 1940s. While sugarcane production reached 37,800 harvested acres in 1949–1950, vegetable farming was the most active segment during that time (Snyder and Davidson 1994). With more flood control in the 1960s and 1970s, sugarcane production increased to 300,000 acres by 1975 (Clarke 1977 as cited by Light and Dineen 1994).

Although a substantial portion of the structural changes for flood control (e.g., perimeter levees and flood control structures) were completed by 1963, subsequently (1965–73) changes had to be made to improve conveyance and meet the environmental water demands of Everglades National Park (ENP; Chimney and Goforth 2001; for review, see Light and Dineen 1994). Construction of the levees had eliminated overland sheet flow southward through the formerly contiguous wetland to ENP.

Initially, Lake Okeechobee was used as a balancing reservoir, as it is today, but also "as a disposal reservoir for natural and artificial drainage of excess storm water from the agricultural lands to the south and east" (Parker et al. 1955). Water from the northern one-third of EAA was routinely back-pumped into Lake Okeechobee (Dickson et al. 1978; Belanger et al. 1989). However, due to mounting concerns regarding the poor quality of this water (stemming from a 1969 USGS report; for review, see Dickson et al. 1978), this practice was stopped in 1979. Instead, all agricultural drainage was directed into the WCAs (Belanger et al. 1989).

The hydrology of the system had thus been fundamentally altered with the initial objective to drain land for human activities (agriculture and urban development).

Later the extensively engineered system of canals, levees, and dikes was regulated with two objectives: (1) minimizing flood risk during the hurricane season and (2) maximizing water storage during the dry season (Light and Dineen 1994).

2.3 Changes in Water Quality

Changes in water quality were first reported in the major canals as early as 1955 by Parker et al. (1955). They found highly mineralized water in the Hillsboro and North New River canals as compared to inflows to Lake Okeechobee. They also found elevated nutrient concentrations in these canals ranging up to 2.4 mg nitrate/L (Parker et al. 1955). They offered several possible explanations for the high concentrations of dissolved minerals in the canals. First, they had found highly mineralized groundwater (i.e., connate seawater) underneath large areas of the Everglades (Parker et al. 1955). They speculated that under low water conditions, i.e., base flow, that a large proportion of water in the canals might originate from this shallow mineralized groundwater (Parker et al. 1955). They also suggested that the quality of water pumped from the agricultural area was another factor possibly responsible for the high dissolved solids in the canals. Alternatively, based on low concentrations observed in Lake Okeechobee and the Everglades, they concluded that flow from these sources were not the source but instead would dilute the canal water (Parker et al. 1955).

The concern was that water in the canals, contaminated with minerals and nutrients, might intrude into the marsh altering its ecology. The difference in water levels between the canal and marsh had been assumed to be the dominant factor controlling canal water intrusion (Surratt et al. 2008). More recently, however, the extent of canal water intrusion into the marsh has been shown to be a bit more complex and a function of the relative inflow and outflow rates for the canal, the duration of the canal inflows as well as sediment elevations and micro-topographic barriers in the marsh (Surratt et al. 2008). Equally important, while the incursion and penetration of chemical constituents carried by surface runoff was a function of the extent of this overland flow, it also depended on the rate of retention of the chemical constituents within the marsh (for review, see Davis 1994). Controversy continues to surround the relative importance of this connate seawater in groundwater versus agricultural drainage as a source of mineralized water to the marshes (for review, see Chap. 2, Volume II). Clearly, connate seawater will have a greater effect on the quality of water in the canals during base flow. Yet, we must also consider how much of this canal water will infiltrate the marshes (and how far it will penetrate) during these low water periods as compared to high water periods where a greater proportion of the canal water will be from stormwater and agricultural drainage.

Follow-up water quality surveys to the canal study were not completed in the marshes until the 1970s (for review, see SFWMD 1992). One of the first surveys was done by McPherson (1976) who sampled 21 sites within WCA 2A and 3A and reported high levels of dissolved solids, trace metals (including copper and zinc), nutrients and pesticides with highest concentrations in the north and lowest in the south. Results of this and other studies indicated "waters draining the agricultural lands to the north of the WCAs was of poor quality…impacting the north and northeast portions of the WCAs" (SFWMD 1992, pg 152). Flora and Rosendahl (1982) reported a 140% increase in conductivity and a 300–400% increase in sodium and chloride concentrations in the waters entering the ENP beginning in the early 1960s.

While the exact timing of nutrient enrichment within the marshes of the WCAs remains uncertain due to the delay in following up the survey by Parker et al. (1955), results from sediment cores would suggest it occurred in the early 1960s upon completion of construction of the water control structures and levels (Craft and Richardson 1993; Bartow et al. 1996). In addition to nutrients, trace metals including copper and zinc were also found to be accumulating in sediments from about this time (Bartow et al. 1996). The authors speculated that the metals probably originated from the EAA since they were known to be applied as micronutrients (Bartow et al. 1996). The timing of sulfur increases in sediment cores from WCA-2A have been found to also correlate with accumulations of phosphorus and probably originated from the EAA (Bates et al. 1998; Chap. 2, Volume II).

2.4 Altered Hydrology and Water Quality Led to Other Changes

Altered hydrology has had many harmful effects on south Florida's environment, some of which were noted as early as 1938 (Beard 1938; Kolipinski and Higer 1969; for review, see Davis and Ogden 1994). Beard (1938), for example, noted alligator holes were drying up, a huge ibis colony site had been abandoned, and snail kites were no longer observed in the area under consideration for establishing ENP. Kolipinski and Higer (1969) later reported more quantifiable impacts such as a decrease in the acreage of wet prairie communities in ENP as a result of shorter periods of inundation. They also reported increased soil oxidation and loss and increased frequency of fires (Kolipinski and Higer 1969; for a review of the implications of a shorter hydroperiod, sediment oxidation and fires on the Hg cycle, see Chap. 1, Volume II). Stieglitz (1962 as cited in SFWMD 1992) reported the first account of cattail (*Typha sp.*) infestations along the Hillsboro canal within WCA-1. Cattails were also replacing sawgrass in WCA-2A with dramatic increases in cattail cover from 1973 through 1990s (Bartow et al. 1996; for review, see Newman et al. 1998).

Agricultural drainage and inputs of phosphorus into this historically oligotrophic system have been considered the primary factor influencing cattail infestations (Koch and Reddy 1992; Newman et al. 1998). Yet, Davis et al. (1994) and Newman et al. (1998) identify a number of variables that influence cattail expansion and the general vegetation patterns found within the Everglades, including climate, sea level, topography, hydroperiod, fire, alligator activity and anthropogenic nutrient inputs. As pointed out by Gunderson (1994), it is important to recognize that the processes affecting these patterns operate at distinct spatial scales (from meters to hundreds of kilometers) and distinct temporal rates (from days to centuries).

The impacts from altered water quantity and quality did not stop with changes in coverage and distribution of emergent vegetation. Because sawgrass detritus was more recalcitrant than cattail detritus, the biological "turnover" or degradation rate was found to be much greater for cattail (Belanger et al. 1989; Davis 1991). This in turn affected oxygen budgets with areas of sawgrass cover generally remaining aerobic while the nutrient–enriched, cattail areas frequently became anaerobic (Belanger et al. 1989). These results have been confirmed in subsequent studies that again show that vegetation regulated aquatic metabolism and dissolved oxygen dynamics, with dense emergent macrophytes limiting dissolved oxygen availability (Hagerthey et al. 2010). This resulted in reduced sediment redox and altered phosphorus biogeochemistry (for review, see Noe et al. 2001; for implications of hypoxia and reducing conditions in pore waters on the Hg cycle the reader is referred to Chaps. 1, 2 and 5, Volume II). Periphyton algae and microbial communities were also exhibiting changes in the Everglades (Reeder and Davis 1983; Swift and Nicholas 1987). Swift and Nicholas (1987), for example, report lower diversity and a shift in species of periphyton at nutrient and mineral enriched sites. Nutrient enrichment was found to shift microbial communities that were responsible for decomposition of leaf litter (Reeder and Davis 1983; for review, see SFWMD 1992). Furthermore, nutrient enrichment, changes in vegetation (as a source of detrital material), altered hydrology (particularly frequency of drying) and sulfate inputs have all likely affected the quantity and quality of DOM in the Everglades (Aiken et al. 2011; for implications on the Hg cycle, see Chap. 4, Volume II). As one example, currently, Everglades DOM is found to be appreciably enriched in sulfur compared to DOM from two other similar large subtropical freshwater wetlands (Chap. 4, Volume II and references therein).

2.5 Summary

As will become evident upon reading the subsequent chapters of this book, the natural Everglades, with its high rainfall and frequent tall convective thunderstorms, its expanse of marshes with shallow, slowly moving water, general lack of freeze-thaw cycles and high average annual temperatures, would have been susceptible to inorganic Hg deposition and its conversion to methylmercury (MeHg). However, the anthropogenic modifications to the Everglades, including altered hydrology, nutrient and sulfate inputs, altered vegetation and microbial communities, changes in DOM quantity and quality (which may have led to changes in photochemistry), lower sediment redox and shifts in biogeochemistry have formed a "perfect storm" making the system extraordinarily efficient at converting inorganic Hg to MeHg. Clearly, the number of biotic and abiotic variables (both internal and external to system, i.e., atmospheric deposition rate) and the spatial scales and temporal rates in which they operate increase the complexity of the Hg problem in the Everglades. This complexity exceeds that which Davis (1994), Gunderson (1994) and others have had to deal with in terms of the spread of cattails. The key to addressing the Hg problem in the Everglades will be to determine what the most influential factors are in leading to high MeHg concentrations in upper trophic level fish and wildlife and to determine which of these factors is most amenable to local control.

References

Aiken GR, Gilmour CC, Krabbenhoft DP, Orem W (2011) Dissolved organic matter in the Florida Everglades: implications for ecosystem restoration. Crit Rev Environ Sci Technol 41:217–248

Bartow SM, Craft CB, Richardson JC (1996) Reconstructing historical changes in Everglades plant community composition using pollen distribution in peat. J Lake Reserv Manag 12:313–322

Bates AL, Spiker EC, Holmes CW (1998) Speciation and isotopic composition of sedimentary sulfur in the Everglades, Florida, USA. Chem Geol 146:155–170

Beard DB (1938). Wildlife reconnaissance, Everglades national park project. Report of the U.S. Department of Interior, National Park Service, Washington, D.C

Belanger TV, Scheidt DJ, Platko JR II (1989) Effects of nutrient enrichment on the Florida Everglades. Lake Reserv Manag 5(1):101–111

Bottcher AB, Izuno FT (1994) Everglades Agricultural Area (EAA) – Water, soil, crop, and environmental management. University Press of Florida, Gainesville, 318 pp

Chimney MJ, Goforth G (2001) Environmental impacts to the Everglades ecosystem: a historical perspective and restoration strategies. Water Sci Technol 44(11–12):93–100

Craft CB, Richardson CJ (1993) Peat accretion and phosphorus accumulation along a eutrophication gradient in the northern Everglades. Biogeochemistry 22:133

Davis JH (1943) The natural features of Southern Florida. The Florida geological survey, bulletin no. 25. Available online at http://ufdc.ufl.edu/UF00000488/00001/1j

Davis SM (1991) Growth, decomposition and nutrient retention of *Cladium jamaicense* Crantz and *Typha domingensis* Pers. in the Florida Everglades. Aquat Bot 40:203–224

Davis SM (1994) Phosphorous inputs and vegetation sensitivity in the Everglades. In: Davis SM, Ogden JC (eds) Everglades: the ecosystem and its restoration. St. Lucie Press, Delray Beach, pp 357–378

Davis SM, Ogden JC (eds) (1994) Everglades: the ecosystem and its restoration. St. Lucie Press, Delray Beach, 826p

Davis SM, Gunderson LH, Park WA, Richardson JR, Mattson JE (1994) Landscape dimension, composition, and function in a changing Everglades ecosystem. In: Davis SM, Ogden JC (eds) Everglades: the ecosystem and its restoration. St. Lucie Press, Delray Beach, pp 419–444

Dickson KG, Federico AC, Lutz JR (1978) Water quality in the Everglades agricultural area and its impact on Lake Okeechobee, Technical publication 78–3. South Florida Water Management District, West Palm Beach

Duever MJ, Meeder JF, Meeder LC, McCollom JM (1994) The climate of South Florida and its role in shaping the Everglades ecosystem. In: Davis SM, Ogden JC (eds) Everglades: the ecosystem and its restoration. St. Lucie Press, Delray Beach, pp 225–248

Finkl CW, Makowski C (2017) The Florida Everglades: an overview of alteration and restoration. In: Finkl C, Makowski C (eds) Coastal wetlands: alteration and remediation. Springer, Dordrecht, pp 3–45. https://doi.org/10.1007/978-3-319-56179-0_1

Flora MD, Rosendahl PC (1982) Historical changes in the conductivity and ionic characteristics of the source water for the shark river slough, Everglades national park, Florida, USA. Hydrobiol 97:249–254

Gleason PJ, Stone P (1994) Age, origin, and landscape evolution of the Everglades peatland. In: Davis SM, Ogden JC (eds) Everglades: the ecosystem and its restoration. St. Lucie Press, Delray Beach, pp 149–197

Gunderson LH (1994) Vegetation: determinants of composition. In: Davis SM, Ogden JC (eds) Everglades: the ecosystem and its restoration. St. Lucie Press, Delray Beach, pp 323–340

Hagerthey SE, Cole JJ, Kilbane D (2010) Aquatic metabolism in the Everglades: the dominance of water column heterotrophy. Limnol Oceanogr 55:653–666

Jones LA (1948) Soils, geology, and water control in the Everglades region, Bulletin no. 442. University of Florida Agricultural Experimental Station, Gainesville

Koch MS, Reddy KR (1992) Distribution of soil and plant nutrients along a trophic gradient in the Florida Everglades. Soil Sci Soc Am J 56:1492–1499

Kolipinski MC, Higer A (1969) Some aspect of the effects of the quantity and quality of the water on biological communities in the Everglades National Park. USGS open field report 69007

Light SS, Dineen JW (1994) Water control in the Everglades: a historical perspective. In: Davis SM, Ogden JC (eds) Everglades: the ecosystem and its restoration. St Lucie Press, Delray Beach, pp 47–84

McCally D (1999) The Everglades: an environmental history. University Press of Florida, Gainesville

McPherson BF, Hendrix GY, Klein H, Tysus HM (1976) The environment of South Florida; a summary report, U.S. geological survey professional paper 1011. United States Government Printing Office, Washington, DC

Morgan KT, McCray JM, Rice RW Gilbert RA, Baucum LE (2015). Review of current sugarcane fertilizer recommendations: a report from the UF/IFAS sugarcane fertilizer standards task force. Institute of Food and Agricultural Sciences (IFAS). Available online at: http://edis.ifas.ufl.edu/ss508

Newman S, Schuette J, Grace JB, Rutchey K, Fontaine T, Reddy KR, Pietrucha M (1998) Factors influencing cattail abundance in the northern Everglades. Aquat Bot 60:265–280

Noe GB, Childers DL, Jones RD (2001) Phosphorus biogeochemistry and the impact of phosphorus enrichment: why is the Everglades so unique? Ecosystems 4:603–624

Parker GG, Ferguson GE, Love SK et al (1955) Water resources of southeastern Florida – with special reference to the geology and ground water of the interior Miami area. US Geol Surv Water-Supply. Paper 1255 157:1–4

Redfield G, VanArman J, Rizzardi K, Chimney M (1999) Chapter 1: introduction to the Everglades interim report. South Florida Water Management District, West Palm Beach. Available online at: http://my.sfwmd.gov/portal/page/portal/pg_grp_sfwmd_sfer/portlet_prevreport/interim-rpt_98/titlepagefinal.pdf

Reeder PB, Davis SM (1983) Decomposition and nutrient uptake and microbial colonization of sawgrass and cattail leaves in water conservation area. 2A, Technical publication 83–4. South Florida Water Management District, West Palm Beach, 24 pp

Snyder GH, Davidson JM (1994) Everglades agriculture: past, present, and future. In: Davis SM, Ogden JC (eds) Everglades: the ecosystem and its restoration. St Lucie Press, Delray Beach, pp 85–115

Solecki WD, Long J, Harwell CC, Myers V, Zubrow E, Ankersen T, Deren C, Feanny C, Hamann R, Hornung L (1999) Human–environment interactions in South Florida's Everglades region: systems of ecological degradation and restoration. Urban Eco 3:305–343

South Florida Water Management District (1992) Surface water improvement and management plan for the Everglades, Supporting information document. South Florida Water Management District, West Palm Beach

Surratt DD, Waldon MG, Harwell MC, Aumen NG (2008) Time series and spatial tracking of polluted canal water intrusion into wetlands of a national wildlife refuge in Florida, USA. Wetlands 28:176–183

Swift DR, Nicholas RB (1987) Periphyton and water quality relationships in the Everglades water conservation areas, Technical Publication 87–2. South Florida Water Management District, West Palm Beach

Chapter 3
Overview of the Atmospheric Mercury Cycle

William M. Landing and Christopher D. Holmes

Abstract Atmospheric deposition is very often the most important source of mercury (Hg) to aquatic systems. Elevated Hg deposition can lead to elevated levels of Hg in aquatic food webs, which degrades the ecosystem and elevates our exposure to Hg via fish consumption. Therefore, an understanding of the atmospheric cycling and deposition of Hg is necessary to develop effective and efficient ways to lower Hg levels in any water body. This is particularly true for the Florida Everglades where direct atmospheric inputs constitute in excess of 95% of the external inputs of Hg to the ecosystem. This chapter thus presents an overview of the atmospheric Hg cycle and discusses key aspects of the cycle that are particularly important to understanding atmospheric Hg deposition in the Florida Everglades.

Keywords Mercury · Atmospheric deposition · Global cycle · Evasion · Legacy mercury · Long-range transport · GEM · GOM · PBM

3.1 Overview of the Atmospheric Hg Cycle

The atmospheric Hg cycle is complex and represents the combination of several input, output and transformation processes that govern the concentrations and dynamics of various forms of Hg in the atmosphere. The forms of Hg in the atmosphere include gaseous elemental Hg (GEM), gaseous oxidized Hg (GOM), and aerosol (particulate-bound) Hg (PBM). GOM can be further subdivided into many neutral chemical species such as gaseous oxides, hydroxides, and halogen complexes with Hg(II). PBM represents all forms of Hg that are associated with particles in the atmosphere, although the majority of the PBM is thought to be in the +2 oxidation state [Hg(II)] Subir et al. 2012).

W. M. Landing (✉) · C. D. Holmes
Department of Earth, Ocean and Atmospheric Science, Florida State University, Tallahassee, FL, USA
e-mail: wlanding@fsu.edu

C. D. Pollman et al. (eds.), *Mercury and the Everglades. A Synthesis and Model for Complex Ecosystem Restoration*,
https://doi.org/10.1007/978-3-030-20070-1_3

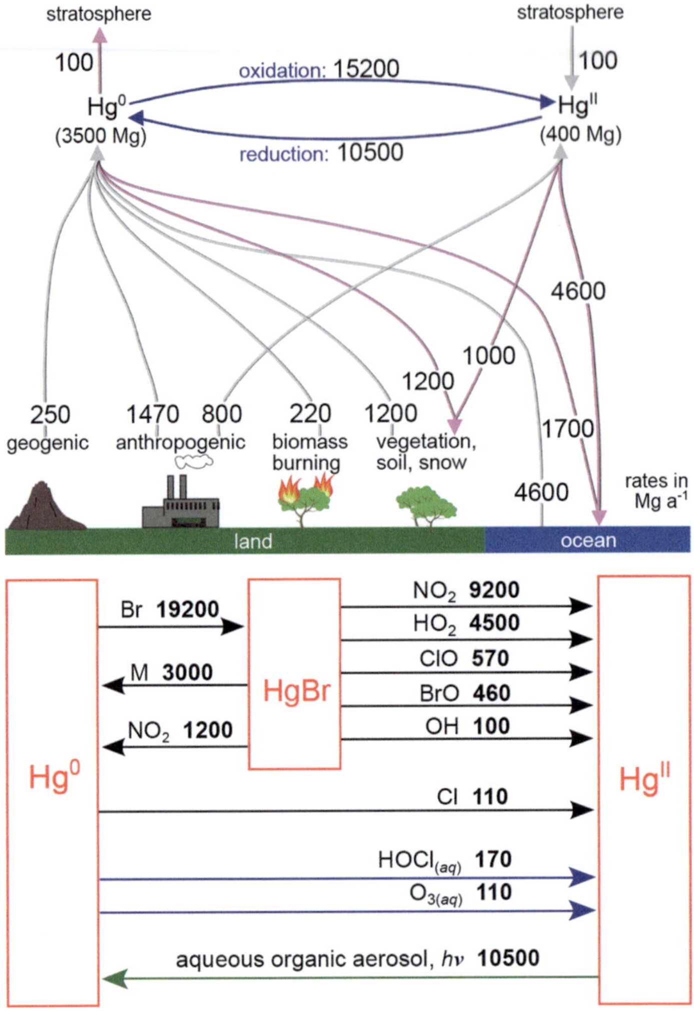

Fig. 3.1 The modern global budget for atmospheric Hg as implemented in the GEOS-Chem model (from Horowitz et al. 2017). The lower panel illustrates the chemical reactions in the atmosphere that cycle atmospheric Hg between GEM [Hg(0)] and Hg(II). Used with permission (https://creativecommons.org/licenses/by/3.0/)

Considering the atmospheric cycle of Hg as it is implemented in the GEOS-Chem model (Fig. 3.1; from Horowitz et al. 2017), GEM accounts for about 90% of the total atmospheric Hg, GOM accounts for about 10%, and PBM accounts for less than a few percent. While GEOS-Chem treats gas-phase and particulate-bound Hg(II) as separate tracers, given the uncertainty in measurement capabilities to separate GOM and PBM (e.g., Gustin et al. 2013) the quantities are combined for model-observation comparisons. It is thought that gaseous oxidized Hg exchanges rapidly with oxidized Hg on aerosols (Amos et al. 2012), and both forms are rapidly deposited by wet and dry deposition processes.

The GEOS-Chem model includes primary emissions of GEM [Hg(0)] and Hg(II) (GOM plus PBM) from geogenic and anthropogenic sources, as well as GEM emissions from biomass-burning, vegetation, soil, snow, and the oceans. Primary natural emissions of Hg (from volcanic activity, crustal degassing, and weathering of geologic deposits) are on the order of 80–600 Mg/yr, primarily as GEM (Wilson et al. 2013). Estimates for current global primary anthropogenic Hg emissions include 2000 Mg/yr (Amos et al. 2013), and 2280 Mg/yr (Zhang et al. 2016). It is estimated that about 35% of the primary anthropogenic Hg emissions are in the form of Hg(II) with 65% as GEM (Zhang et al. 2016; Streets et al. 2017). Re-emission of legacy anthropogenic Hg (mostly as GEM) is estimated at around 4700 Mg/yr.

As described by Horowitz et al. (2017), in the gas phase, GEM is oxidized to Hg(II) in a two-step process involving halogen radicals (e.g., Br, Cl) or OH in the first step and a variety of oxidants in the second step (NO_2, HO_2, ClO, BrO, OH, Cl, Br). In the aqueous phase (in cloud water, rain, and fog), GEM is oxidized by oxidants including aqueous HOCl, OH, and ozone. Gas phase oxidation of GEM by ozone may also be important and is an area of current study. Photochemical reduction of atmospheric Hg(II) back to GEM occurs through photolysis in air and in aqueous reactions in clouds and aerosol (Horowitz et al. 2017; Saiz-Lopez et al. 2018).

The concentrations of most of these potential GEM oxidants are well known from decades of atmospheric chemistry measurements and modeling (e.g. Seinfeld and Pandis 2016). Recent advances in techniques for measuring halogens, however, have revealed higher concentrations of reactive bromine than were expected in the troposphere (Wang et al. 2015; Coburn et al. 2016). Ongoing field experiments continue to map the abundance of reactive halogens, but these high concentrations can likely be explained by emissions of biogenic halocarbons from marine algae, debromination of sea-salt aerosols, and recycling of background aerosols (Sherwen et al. 2016).

Hg deposition occurs via wet and dry deposition processes to the land (vegetation, soil, snow) and the oceans. It is estimated that about 65–80% of total Hg deposition on a global scale is due to wet and dry deposition of Hg(II), while 20–35% is due to dry deposition of GEM. Hg(II) in the lower atmosphere (the boundary layer) will adsorb (dry deposit) to vegetation and open water and be taken up by sea salt aerosols. In GEOS-Chem, on a global basis, dry deposition accounts for about 40% of total Hg(II) deposition while wet deposition of Hg(II) accounts for about 60% (Horowitz et al. 2017). Wet deposition of Hg occurs when cloud water absorbs Hg(II) and when precipitation scavenges aerosol Hg. Most of the wet deposition is concentrated in the tropics, where rain is most abundant. GEM is not a significant contributor to wet deposition due to its relatively low solubility in water.

It is essential to accurately quantify all emission sources of both GEM and Hg(II), and the potentially rapid redox cycling in the atmosphere between these forms, since deposition of Hg(II) will have significant impacts on ecosystem bioaccumulation of Hg on local and regional scales. This is critically important since ecosystem studies have shown that recently deposited Hg(II) is taken up into aquatic food webs most quickly (Branfireun et al. 2005; see Section 3.2 in Chap. 3, Volume III for a more detailed discussion).

Using the GEM and Hg(II) emission rates in Fig. 3.1, one can estimate that the atmospheric residence time of GEM is on the order of 6 months (reservoir size of 3500 Mg divided by the total input rate of 7740 Mg/yr). Because of its low solubility in water and its relatively long atmospheric residence time, GEM emissions are transported far from their sources and become more well-mixed in the atmosphere.

The residence time of Hg(II) is on the order of 25 days (400 Mg divided by net input of 5600 Mg/yr). The oxidized forms of atmospheric Hg (GOM and PBM) are very soluble in water, are easily scavenged by rainfall, and adsorb to surfaces, resulting in deposition much closer to their sources.

Figure 3.2a (from AMAP/UNEP 2015) shows the emissions of Hg on regional and local scales. High Hg emission rates are associated with population density and with artisanal-scale gold mining (ASGM) in South America and central and southern Africa. On a global basis, over 90% of modern Hg emission is as GEM, although GOM and PBM can account for up to 20% in certain areas (southeast Asia, Europe, and North America). Figure 3.2b shows that the modeled concentrations of GEM are highest near regions of high population and high anthropogenic emissions (China, India, Europe) and in South America and Africa due to ASGM. Atmospheric Hg concentrations are significantly lower in the southern hemisphere due to the lower population and the slow exchange of air across the tropics. Figure 3.2c shows model results for the global deposition of atmospheric Hg (which includes GEM, GOM, and PBM). The deposition patterns reveal several important factors that influence atmospheric Hg deposition. Uptake of the GEM by forested areas yields high deposition across South America and central and southern Africa, although much of that GEM is re-emitted on an annual basis. Elevated Hg deposition associated with high precipitation rates is seen along the Inter-Tropical Convergence Zones (ITCZ) in the equatorial Atlantic, Pacific, and Indian Oceans and in the Southern Ocean near Antarctica. This is interpreted as being due to more efficient oxidation of GEM to Hg(II) species which are easily scavenged during precipitation events. Combinations of regionally elevated Hg emissions and elevated rainfall yield higher Hg deposition in the southeastern US and in the western North Pacific and western North Atlantic. The highest Hg deposition is associated with high population density and high anthropogenic emissions as seen in Europe, India, and China.

As discussed above, the residence time of GEM in the atmosphere is relatively long (6–12 months), while Hg(II) has a much shorter residence time (days to weeks). As a result, GEM from all emission sources in the northern hemisphere is relatively well-mixed while Hg(II) shows strong local and regional concentration differences. Hg(II) is also quickly removed from the atmosphere by precipitation while GEM is not. Thus, local and regional sources of Hg(II) will have greater local and regional impacts on Hg deposition.

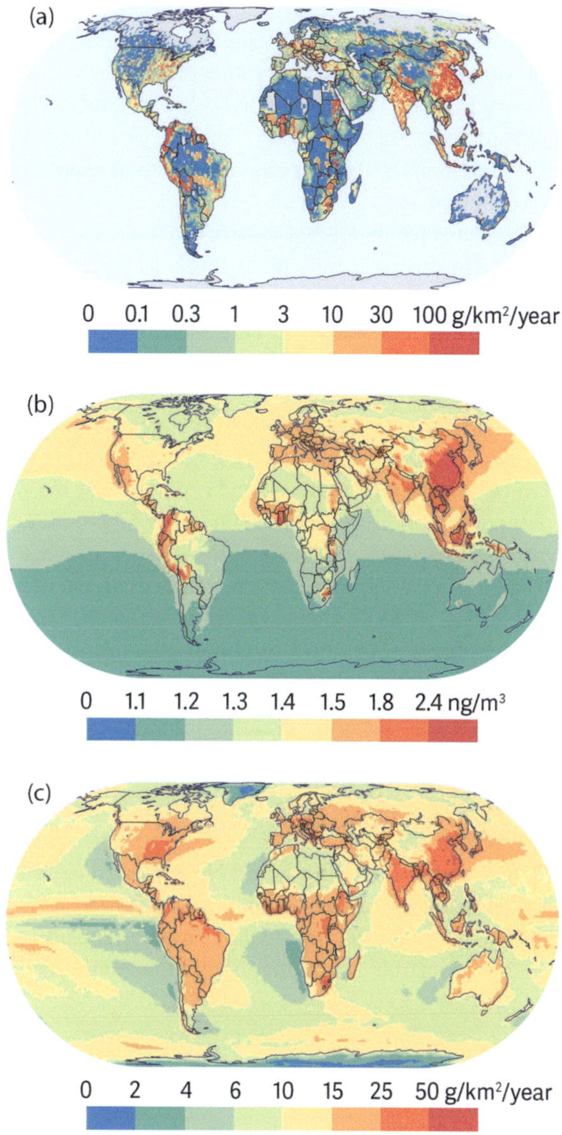

Fig. 3.2 (**a**) Emission of Hg to the atmosphere. (**b**) Modeled concentrations of GEM in the atmosphere in 2013. (**c**) Modeled annual Hg deposition in 2013 (from AMAP/UNEP 2015). Based on an ensemble of three chemical transport models (GLEMOS, GEOS-Chem, and GMHG). Used with permission from AMAP/UNEP (2015)

Except for a few areas where industrial effluent containing Hg contaminates the local environment, the dominant source of Hg to the global environment is from atmospheric deposition. Atmospheric deposition is the main source of Hg to Florida, especially the Florida Everglades (USEPA 1996). Modeling studies also show that the majority (>98%) of the atmospherically Hg deposited in Florida is from Hg sources outside of Florida (USEPA 2008; FDEP 2013). The model results for Hg deposition in the Everglades are reviewed in Chap. 5, this volume and Chap. 3, Volume III.

Amos et al. (2013) estimated that about 13% of the atmospheric Hg deposited globally was from natural sources. According to their calculations, the current atmospheric deposition of Hg on a global basis can be apportioned as follows: 3% primary natural; 10% legacy natural, 27% primary anthropogenic, and 60% legacy anthropogenic Hg. Variations from this global Hg deposition apportionment are to be expected on local and regional scales where anthropogenic emissions of oxidized Hg are important.

Estimates for the global budget of Hg (Table 3.1) suggest that the modern atmospheric Hg reservoir may be as much as 7.5 times larger relative to natural levels due to increasing anthropogenic Hg emissions since the start of the industrial revolution (Amos et al. 2013). This conclusion was questioned by Engstrom et al. (2014); a detailed study of Hg in lake sediment cores indicated that the magnitude of early Hg emissions from artisanal gold mining were overestimated by Amos et al. (2013), and that the modern/natural deposition ratio should be closer to 3–5. Amos et al. (2015) tried to harmonize the time periods for observed and modeled estimates of anthropogenic enrichment and reported that the mean enrichment from the pre-colonial era (3000 BC–1550 AD) to the twentieth century maximum (1950–1975) was 19 for varved sediment cores and 26 for peat cores. Consistent with these results, Enrico et al. (2017) estimated enrichment of 15 ± 4 from the Holocene to the present-day as reconstructed from a peat bog. Regardless of the exact value, Hg concentrations in the modern atmosphere, and atmospheric Hg deposition, are significantly elevated over pre-industrial times.

Emissions of Hg to the atmosphere from anthropogenic activities over time since the 1400s were summarized by Beal et al. (2015; Fig. 3.3). Primary anthropogenic Hg emissions were slightly higher (~500 Mg/yr) from 1450–1650, then declined to ~100–150 Mg/yr from 1650–1850. Starting around 1850, Hg emissions increased dramatically, associated with the American gold and silver rush period when amalgamation with liquid Hg was used to extract precious metals from ores. Emissions of Hg peaked again from 1940–1975 due to increased commercial and industrial use. Temporary declines in Hg deposition following those two emission peaks demonstrate the relatively rapid response time of atmospheric Hg deposition

Table 3.1 Estimated amounts of Hg in the atmosphere, terrestrial, and oceanic reservoirs for natural, preindustrial (pre-1840), and modern conditions, with enrichment factors for the modern reservoir relative to natural levels

Compartment		Natural	Preindustrial	Present day	Modern/Natural ratio
Atmosphere		700	2000	5300	7.5
Terrestrial	Fast	1900	4800	11,000	5.8
	Slow	9800	24,000	50,000	5.1
	Armored	170,000	180,000	210,000	1.2
Ocean	Surface	530	1300	3100	5.9
	Subsurface	26,000	67,000	140,000	5.3
	Deep	95,000	130,000	200,000	2.1

Adapted from Amos et al. (2013)
Units are Mg of Hg

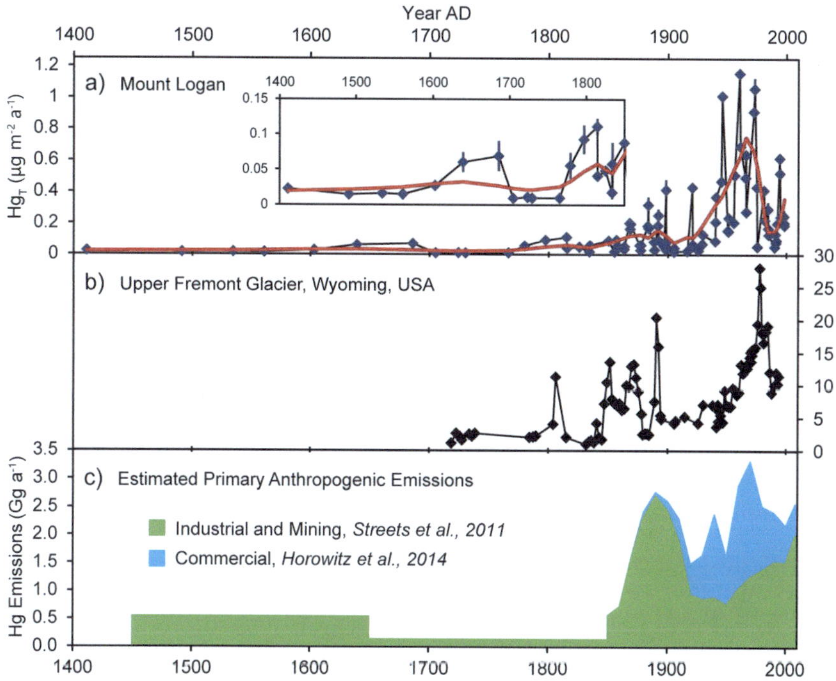

Fig. 3.3 Records of total Hg deposition in mountain glacier ice cores from Alaska (panel **a**) and Wyoming (panel **b**) with estimates for primary anthropogenic Hg emissions (panel **c**). (Beal et al. 2015). The red lines in panel **a** are smoothed estimates obtained from locally weighted polynomial regression (LOESS). (Reprinted with permission from Beal et al. (2015). Copyright 2015 American Chemical Society)

to changes in the atmospheric loading. Primary anthropogenic input has apparently declined by about 20% from 1990 to 2010 due to reductions in the use of Hg in man-made products (and the incineration of those products) and improved Hg scrubbing during coal combustion (Zhang et al. 2016).

Estimates for the global distribution of atmospheric Hg deposition rates (subdivided by source region) were derived from an ensemble of three chemical transport models (GLEMOS, GEOS-Chem, GMHG) (Fig. 3.4; AMAP/UNEP 2015). The model results demonstrate strong seasonality for Hg deposition and the importance of natural and legacy Hg emissions. For example, over North America, Hg deposition peaks in the spring and summer months. Emission of Hg from North American sources only account for 5–12% of the total Hg deposition over North America (lowest during the spring and summer months). Legacy anthropogenic and natural emissions and primary natural emissions dominate the deposition of atmospheric Hg everywhere but are least important (60–70%) over East Asia and South Asia where local and regional emissions are more dominant. Further research on both modern and historical Hg deposition rates on a global scale is needed to help constrain these model estimates.

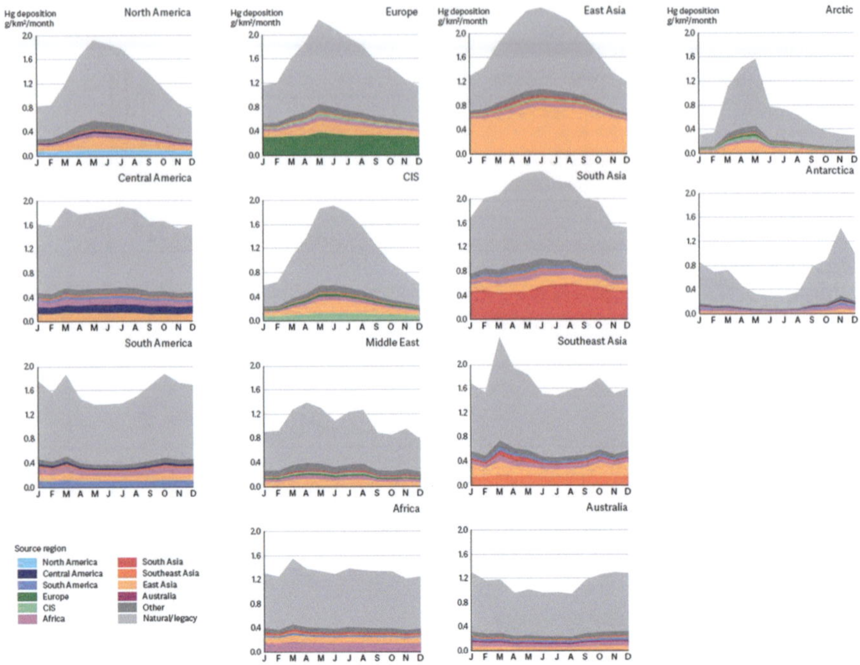

Fig. 3.4 Monthly Hg deposition (2013) attributed to Hg emissions from various geographic regions. (Used with permission from AMAP/UNEP 2015)

3.2 Importance of Atmospheric Hg Speciation on Transport and Deposition

As noted above, gaseous elemental mercury (GEM) has a lifetime in the atmosphere of approximately 6–12 months. Species with such long atmospheric lifetimes are usually widely dispersed throughout the troposphere with relatively modest variations in concentration (Fig. 3.2b). The resulting dry deposition of GEM is therefore driven primarily by global sources as opposed to local emissions.

Oxidized mercury [Hg(II)], in contrast, has a lifetime of only a few weeks in the free troposphere (~2–18 km altitude), like other aerosols (e.g. sulfate) and highly soluble gases (e.g. nitric acid) (Seinfeld and Pandis 2016). However, within the planetary boundary layer (<2 km altitude), both GOM and PBM have much shorter lifetimes, around 1 day, since they are removed by dry deposition on contact with the surface. Consequently, oxidized mercury is predominantly deposited near its emission source (up to 100 km away), although it can be transported much farther if it is lifted above the boundary layer.

While very high Hg(II) concentrations have been reported around anthropogenic sources, observations from aircraft sampling campaigns and monitoring on mountain tops show that Hg(II) concentrations generally increase with altitude. This is believed to be caused by oxidation of GEM to produce Hg(II) that accumulates in the upper troposphere and is consistent with the model results shown in Fig. 3.1 where net oxidation of GEM to Hg(II) is on the order of 5000 Mg/yr. This region of the atmosphere rarely experiences rain and it can take a month or longer for air to mix down to lower altitudes. As a result, Hg(II) deposition is high where large-scale high-pressure systems transport air downwards (Shah and Jaegle 2017). This happens mainly at sub-tropical latitudes (which includes Florida). In addition, Hg(II) deposition is high in regions with frequent deep convection, like Florida, that can scavenge Hg(II) directly from high altitudes.

3.3 Modifiers to the Atmospheric Hg Cycle that Influence Hg Deposition in the Everglades

Guentzel et al. (2001) reported rainfall Hg data collected across the state of Florida from 1993 through 1996 (the Florida Atmospheric Mercury Study; FAMS) and concluded that strong convective thunderstorms in the summer months were responsible for ~84% of the annual rainfall Hg deposition vs. lower altitude rain events and winter-time frontal storms. They showed that wet deposition of Hg was nearly uniform at 20–23 μg/m^2-yr from West Palm Beach to Ft. Myers and throughout the Florida Everglades and that rainfall Hg deposition reached a peak during the summer months when convective thunderstorms are very common over southern Florida. Similar results were reported by Shanley et al. (2015) who observed elevated rainfall Hg concentrations and fluxes from summertime thunderstorms at a pristine site in Puerto Rico. To explain their findings, Guentzel et al. (2001) hypothesized that there must be a rapidly-replenished reservoir of "background" Hg(II) in the free troposphere that supplies Hg(II) to these thunderstorms when the southeasterly trade winds shift north and flow across southern Florida.

When Guentzel et al. (2001) obtained their rainfall Hg data across southern Florida (1993–1996), no data had been published on Hg(II) concentrations in the free troposphere anywhere in the world, and the hypothesis could not be tested. Since then, aircraft sampling for Hg(II) in the free troposphere over the Bahamas found very high Hg(II) concentrations (50–250 pg/m^3) between 1–4 km height (Landis et al. 2005; Sillman et al., 2007). Brooks et al. (2014) reported Hg(II) concentrations as high as 150 pg/m^3 at 4 km during the spring and summer months over Tullahoma, TN (10–30 times higher than concurrent ground level measurements). Both aircraft Hg(II) data sets were collected using KCl-coated annular denuders that subsequently have been reported to under-collect Hg(II) by as much as a factor of 2 (Swartzendruber et al. 2009; Gustin et al. 2013); thus, the concentrations of Hg(II) in the free troposphere may be significantly higher than were reported. This reservoir of Hg(II) could easily support the summertime rainfall Hg concentrations and fluxes that Guentzel et al. (2001) reported across southern Florida. Further support for this hypothesis comes from Holmes et al. (2016) who used rainfall Hg data from the NADP-MDN program across the continental US and GridSat-B1 satellite cloud-top brightness data to conclude that thunderstorms generate 50% higher rainfall Hg concentrations relative to weak convective or stratiform events of equal precipitation depth. In a modeling study, Kaulfus et al. (2017) concluded that, after controlling for precipitation depth, the highest Hg deposition occurs in supercell thunderstorms, with decreasing deposition in disorganized thunderstorms, quasi-linear convective systems (QLCS), extratropical cyclones, light rain, and land-falling tropical cyclones. Convective morphologies (supercells, disorganized, and QLCS) enhanced wet deposition by a factor of at least 1.6 relative to non-convective morphologies.

Lacking any data on Hg(II) concentrations in the free troposphere, Guentzel et al. (2001) likely overestimated that local and regional emissions of Hg could account for 30–46% of the rainfall Hg deposition across southern Florida. More recent estimates for the importance of local and regional source of atmospheric Hg for Hg deposition in the Everglades are discussed in Chap. 5, this volume. Briefly, the Florida Mercury TDML Study (FDEP 2013) concluded that Florida sources accounted for less than 2% of the total Hg deposition at the Everglades National Park site.

3.4 Evasion of Hg

Recent modeling studies of the global Hg cycle suggest that uptake and evasion losses of GEM are roughly balanced for terrestrial ecosystems (Amos et al. 2013; Song et al. 2015; Horowitz et al. 2017; see Fig. 3.1). The very low GEM evasion estimate of Amos et al. (2014) was an exception to these other three studies, with evasion loss approximately sixfold less, although their uptake fluxes are similar. Excluding the very low evasion flux indicated by Amos et al. (2014), the global modeling studies cited above indicate that GEM evasion accounts for 50 ± 5% of the total Hg deposition to terrestrial surfaces (wet plus dry GEM, GOM, and PBM). In a more detailed model of Hg uptake by terrestrial ecosystems, Smith-Downey et al. (2010) suggest that plants take up GEM but do not release 100% of it. Plants also collect GOM and PBM, some of which is photo-reduced back to GEM. Overall, GEM evasion accounted for 67% of the total Hg deposition to terrestrial systems (Smith-Downey et al. 2010). In a field experiment, Poulain et al. (2006) measured 33–59% GEM evasion from mesocosms spiked with Hg(II). These various estimates for GEM evasion agree reasonably well and suggest that GEM evasion from wetland ecosystems is probably on the order of 50–60% of total Hg deposition.

With respect to the oceans, however, current estimates indicate that evasion losses of GEM from the ocean surface greatly exceed uptake fluxes of GEM from the atmosphere (median ratio for the four studies is 3:1). The large GEM evasion losses from the surface ocean are due to Hg(II) reduction in surface waters. The

GEM flux approximately balances the deposition of GOM and PBM and influences how rapidly the Hg reservoir in the surface ocean turns over (~ 0.25 year; Amos et al. 2013; Song et al. 2015).

3.5 Conclusions

The atmospheric Hg cycle is extremely complex in part due to the variety of emission sources and emission speciation profiles (GEM vs. Hg(II)) and their geographic distribution. It is also complex because transformations of GEM to GOM are driven by a variety of atmospheric oxidants with different sources and distributions in the atmosphere and because the physics of atmospheric mixing varies spatially. Significant progress has been made in recent years on modeling the in situ processes influencing GEM oxidation and Hg(II) reduction, but there is still much to learn. The residence times for GEM and Hg(II) in the troposphere with respect to their inputs are on the order of 6–12 months and 25 days, respectively, while Hg(II) in the boundary layer is deposited much more quickly than GEM. As a result, the atmospheric deposition of Hg is strongly influenced by the magnitude and speciation of Hg emissions on local and regional scales, and by the patterns of vegetation and rainfall across the globe. Both field data and model simulations suggest that convective thunderstorms yield significantly higher rainfall Hg concentrations and fluxes due to scavenging of Hg(II) from the free troposphere. Local and regional Hg emissions can have significant impacts on regional Hg deposition; the lack of accurate emission inventories and emission profiles [GEM vs. Hg(II)] however continues to affect the reliability of model predictions for Hg deposition.

The state of knowledge on the atmospheric mercury cycle continues to advance, and more recent information on Hg emissions and deposition has been summarized in the 2018 UNEP Global Mercury Assessment (GMA).

Acknowledgments We would like to thank Curt Pollman and Hannah Horowitz for their very helpful reviews of this chapter.

References

AMAP/UNEP (2015) Global mercury modelling: update of modelling results in the global mercury assessment 2013. Arctic Monitoring and Assessment Programme/UNEP Chemicals Branch, Oslo/Geneva. iv + 32 pp

Amos HM, Jacob DJ, Holmes CD, Fisher JA, Wang Q, Yantosca RM, Corbitt ES, Galarneau E, Rutter AP, Gustin MS, Steffen A, Schauer JJ, Graydon JA, St Louis VL, Talbot RW, Edgerton ES, Zhang Y, Sunderland EM (2012) Gas-particle partitioning of atmospheric Hg(II) and its effect on global mercury deposition. Atmos Chem Phys 12:591–603

Amos HM, Jacob DJ, Streets DG, Sunderland EM (2013) Legacy impacts of all-time anthropogenic emissions on the global mercury cycle. Glob Biogeochem Cycles 27:410–421
Amos HM, Jacob DJ, Kocman D, Horowitz HM, Zhang YX, Dutkiewicz S, Horvat M, Corbitt ES, Krabbenhoft DP, Sunderland EM (2014) Global biogeochemical implications of mercury discharges from rivers and sediment burial. Environ Sci Technol 48:9514–9522
Amos HM, Sonke JE, Obrist D, Robins N, Hagan N, Horowitz HM, Mason RP, Witt M, Hedgecock IM, Corbitt ES, Sunderland EM (2015) Observational and modeling constraints on global anthropogenic enrichment of mercury. Environ Sci Technol 49:4036–4047
Beal SA, Osterberg EC, Zdanowicz CM, Fisher DA (2015) Ice core perspective on mercury pollution during the past 600 years. Environ Sci Technol 49:7641–7647
Branfireun BA, Krabbenhoft DP, Hintelmann H, Hunt RJ, Hurley JP, Rudd JWM (2005) Speciation and transport of newly deposited mercury in a boreal forest wetland: a stable mercury isotope approach. Water Resour Res 41:11
Brooks S, Ren XR, Cohen M, Luke WT, Kelley P, Artz R, Hynes A, Landing W, Martos B (2014) Airborne vertical profiling of mercury speciation near Tullahoma, TN, USA. Atmosphere 5:557–574
Coburn S, Dix B, Edgerton E, Holmes CD, Kinnison D, Liang Q, ter Schure A, Wang SY, Volkamer R (2016) Mercury oxidation from bromine chemistry in the free troposphere over the southeastern US. Atmos Chem Phys 16:3743–3760
Engstrom DR, Fitzgerald WF, Cooke CA, Lamborg CH, Drevnick PE, Swain EB, Balogh SJ, Balcom PH (2014) Atmospheric Hg emissions from preindustrial gold and silver extraction in the Americas: a reevaluation from lake-sediment archives. Environ Sci Technol 48:6533–6543
Enrico M, Le Roux G, Heimburger LE, Van Beek P, Souhaut M, Chmeleff J, Sonke JE (2017) Holocene atmospheric mercury levels reconstructed from peat bog mercury stable isotopes. Environ Sci Technol 51:5899–5906
Florida Department of Environmental Protection (2013) Mercury TMDL for the state of Florida, Final report, watershed evaluation and TMDL section, October 24, 2013. https://floridadep.gov/sites/default/files/Mercury-TMDL.pdf
Guentzel JL, Landing WM, Gill GA, Pollman CD (2001) Processes influencing rainfall deposition of mercury in Florida. Environ Sci Technol 35:863–873
Gustin MS, Huang JY, Miller MB, Peterson C, Jaffe DA, Ambrose J, Finley BD, Lyman SN, Call K, Talbot R, Feddersen D, Mao HT, Lindberg SE (2013) Do we understand what the mercury speciation instruments are actually measuring? Results of RAMIX. Environ Sci Technol 47:7295–7306
Holmes CD, Krishnamurthy NP, Caffrey JM, Landing WM, Edgerton ES, Knapp KR, Nair US (2016) Thunderstorms increase mercury wet deposition. Environ Sci Technol 50:9343–9350
Horowitz HM, Jacob DJ, Zhang YX, Dibble TS, Slemr F, Amos HM, Schmidt JA, Corbitt ES, Marais EA, Sunderland EM (2017) A new mechanism for atmospheric mercury redox chemistry: implications for the global mercury budget. Atmos Chem Phys 17:6353–6371
Kaulfus AS, Nair U, Holmes CD, Landing WM (2017) Mercury wet scavenging and deposition differences by precipitation type. Environ Sci Technol 51:2628–2634
Landis MS, Lynam M, Stevens RK (2005) The monitoring and modeling of mercury species in support of local regional and global modeling. In: Pirrone N, Mahaffey KR (eds) Dynamics of mercury pollution on regional and global scales. Kluwer Academic, New York, pp 123–151
Poulain AJ, Orihel DM, Amyot M, Paterson MJ, Hintelmann H, Southworth GR (2006) Relationship to aquatic between the loading rate of inorganic mercury ecosystems and dissolved gaseous mercury production and evasion. Chemosphere 65:2199–2207
Saiz-Lopez A, Sitkiewicz SP, Roca-Sanjuán D, Oliva-Enrich JM, Dávalos JZ, Notario R et al (2018) Photoreduction of gaseous oxidized mercury changes global atmospheric mercury speciation, transport and deposition. Nat Commun 9(1):1–9. https://doi.org/10.1038/s41467-018-07075-3
Seinfeld JH, Pandis SN (2016) Atmospheric chemistry and physics: from air pollution to climate change, 3rd edn. Wiley, New York. Published 2016, ISBN-10: 1118947401

Shah V, Jaegle L (2017) Subtropical subsidence and surface deposition of oxidized mercury produced in the free troposphere. Atmos Chem Phys 17:8999–9017
Shanley JB, Engle MA, Scholl M, Krabbenhoft DP, Brunette R, Olson ML, Conroy ME (2015) High mercury wet deposition at a "clean air" site in Puerto Rico. Environ Sci Technol 49:12474–12482
Sherwen T, Schmidt JA, Evans MJ, Carpenter LJ, Grossmann K, Eastham SD, Jacob DJ, Dix B, Koenig TK, Sinreich R, Ortega I, Volkamer R, Saiz-Lopez A, Prados-Roman C, Mahajan AS, Ordonez C (2016) Global impacts of tropospheric halogens (Cl, Br, I) on oxidants and composition in GEOS-Chem. Atmos Chem Phys 16:12239–12271
Sillman S, Marsik FJ, Al-Wali KI, Keeler GJ, Landis MS (2007) Reactive mercury in the troposphere: model formation and results for Florida, the Northeastern United States, and the Atlantic Ocean. J Geophys Res-Atmos 112:17
Smith-Downey NV, Sunderland EM, Jacob DJ (2010) Anthropogenic impacts on global storage and emissions of mercury from terrestrial soils: insights from a new global model. J Geophys Res Biogeosci 115:11
Song S, Selin NE, Soerensen AL, Angot H, Artz R, Brooks S, Brunke EG, Conley G, Dommergue A, Ebinghaus R, Holsen TM, Jaffe DA, Kang S, Kelley P, Luke WT, Magand O, Marumoto K, Pfaffhuber KA, Ren X, Sheu GR, Slemr F, Warneke T, Weigelt A, Weiss-Penzias P, Wip DC, Zhang Q (2015) Top-down constraints on atmospheric mercury emissions and implications for global biogeochemical cycling. Atmos Chem Phys 15:7103–7125
Streets DG, Horowitz HM, Jacob D, Lu ZF, Levin L, ter Schure AFH, Sunderland EM (2017) Total mercury released to the environment by human activities. Environ Sci Technol 51:5969–5977
Subir M, Ariya PA, Dastoor AP (2012) A review of the sources of uncertainties in atmospheric mercury modeling II. Mercury surface and heterogeneous chemistry – a missing link. Atmos Environ 46:1–10
Swartzendruber PC, Jaffe DA, Finley B (2009) Development and first results of an aircraft-based, high time resolution technique for gaseous elemental and reactive (oxidized) gaseous mercury. Environ Sci Technol 43:7484–7489
UN Environment (2019) Global mercury assessment 2018. UN Environment Programme, Chemicals and Health Branch, Geneva
United States Environmental Protection Agency (USEPA) (1996) South Florida ecosystem assessment interim report. Monitoring for adaptive management: implications for ecosystem restoration. Region 4 SESD and ORD. EPA-904-96-008. December 1996
USEPA REMSAD (2008) Model-based analysis and tracking of airborne mercury emissions to assist in watershed planning. Watershed branch (4503-T), Office of Wetlands, Oceans, and Watersheds, U.S. Environmental Protection Agency, 1200 Pennsylvania Avenue, NW, Washington, DC 20460 (2009)
Wang SY, Schmidt JA, Baidar S, Coburn S, Dix B, Koenig TK, Apel E, Bowdalo D, Campos TL, Eloranta E, Evans MJ, DiGangi JP, Zondlo MA, Gao RS, Haggerty JA, Hall SR, Hornbrook RS, Jacob D, Morley B, Pierce B, Reeves M, Romashkin P, ter Schure A, Volkamer R (2015) Active and widespread halogen chemistry in the tropical and subtropical free troposphere. Proc Natl Acad Sci U S A 112:9281–9286
Wilson S, Kindbom K, Yaramenka K, Steenhuisen F, Telmer K, Munthe J (2013) Global emissions of mercury to the atmosphere, in technical background report for the global mercury assessment 2013. Arctic Monitoring and Assessment Programme, Geneva, pp 4–37
Zhang YX, Jacob DJ, Horowitz HM, Chen L, Amos HM, Krabbenhoft DP, Slemr F, St Louis VL, Sunderland EM (2016) Observed decrease in atmospheric mercury explained by global decline in anthropogenic emissions. Proc Natl Acad Sci U S A 113:526–531

Chapter 4
Atmospheric Deposition Flux of Mercury to the Everglades

Krish Vijayaraghavan and Curtis D. Pollman

Abstract This chapter presents a summary of data on atmospheric mercury (Hg) deposition under precipitating and non-precipitating conditions within and near the Everglades from monitoring data and other estimates and a comparison with data from elsewhere in the United States. Hg deposition to the Everglades is among the highest in the country and has continued to be so over the past two decades.

Keywords Mercury · Wet deposition · Dry deposition · Spatial variability · Oxidation states · Temporal trend

4.1 Introduction

The deposition of inorganic mercury (Hg) from the atmosphere to the Earth's surface represents a significant fraction (over 95%) of the external loading of Hg to the Florida Everglades (USEPA 1996; Guentzel et al. 1998; FDEP 2003, 2013). Atmospheric deposition may occur under both wet and dry conditions and is influenced by precipitation patterns and other meteorological conditions, the presence of emission sources, local terrain and land cover, atmospheric concentrations and transformations of Hg, and long-range atmospheric transport of Hg. As discussed in Chap. 3 (this volume), Hg in both atmospheric emissions and the ambient atmosphere can be present in a variety of oxidation states and phases, which in turn exhibit varying deposition characteristics. Due to its limited solubility in water, Hg(0) (gaseous elemental Hg or GEM) has negligible wet deposition. The wet deposition of Hg primarily consists of gaseous divalent Hg(II) (gaseous oxidized

K. Vijayaraghavan (✉)
Ramboll US Corporation, Novato, CA, USA
e-mail: kvijay@ramboll.com

C. D. Pollman
Aqua Lux Lucis, Inc, Gainesville, FL, USA
e-mail: cpollman@aqualuxlucis.org

C. D. Pollman et al. (eds.), *Mercury and the Everglades. A Synthesis and Model for Complex Ecosystem Restoration*,
https://doi.org/10.1007/978-3-030-20070-1_4

Hg or GOM) which is highly soluble in water, with a smaller fraction of particulate-bound divalent Hg [PBM, also referred to as Hg(p)]. GEM undergoes dry deposition under non-precipitating conditions but a portion is re-emitted back to the atmosphere. GOM dry deposits rapidly, especially near emission sources, as does coarse PBM, while fine particulate PBM deposits more slowly. These chemical forms of Hg also inter-convert as discussed in Chap. 3 (this volume) and thus the deposition of Hg is dependent on emissions and concentrations of all forms of Hg. While most atmospheric deposition of Hg is due to the inorganic chemical forms discussed above, there is also a relatively small fraction of methylmercury (MeHg) present in wet deposition.

The atmospheric deposition flux of Hg in southern Florida is among the highest in the United States (see next section). Several measurement campaigns and other studies have increased our knowledge of the deposition of atmospheric Hg to the Everglades. The findings from these networks and studies are discussed below in the context of wet and dry deposition flux of Hg to the Everglades together with a comparison to deposition in other geographic regions, in Florida and elsewhere in the United States.

4.2 Wet Deposition

Hg in precipitation has been measured at monitoring stations in the Mercury Deposition Network (MDN) in North America for over 20 years. A component of the National Atmospheric Deposition Program (NADP), the MDN is the only network providing a long-term record of Hg rain concentration and wet deposition in the U.S. and Canada (NADP 2017). All MDN sites follow standard procedures and have uniform precipitation chemistry collectors and gauges with weekly samples. The network has over 100 monitoring stations in the U.S and Canada (Fig. 4.1) with five stations in Florida active in 2016 including the following three located within or near the Everglades:

- Everglades National Park – Beard Research Center (MDN site ID FL11), active since March 1996 and operated by the Everglades National Park.
- Everglades – Western Broward County (MDN site ID FL97), active since November 2006 and operated by the South Florida Water Management District.
- Everglades – South Palm Beach County (MDN site ID FL95), active since April 2015 and operated by the South Florida Water Management District.

The measured annual Hg wet deposition fluxes in 2016 at MDN stations across the U.S. and Canada are shown in Fig. 4.2. The map contour surface represents a gridded interpolation developed by the NADP (2017) using discrete MDN wet deposition measurements, terrain, a precipitation grid derived from NADP precipitation observations, and a national precipitation grid developed by the PRISM (Parameter-elevation Regressions on Independent Slopes Model) Climate Group (PRISM 2017). The three stations near the Everglades experience much higher Hg

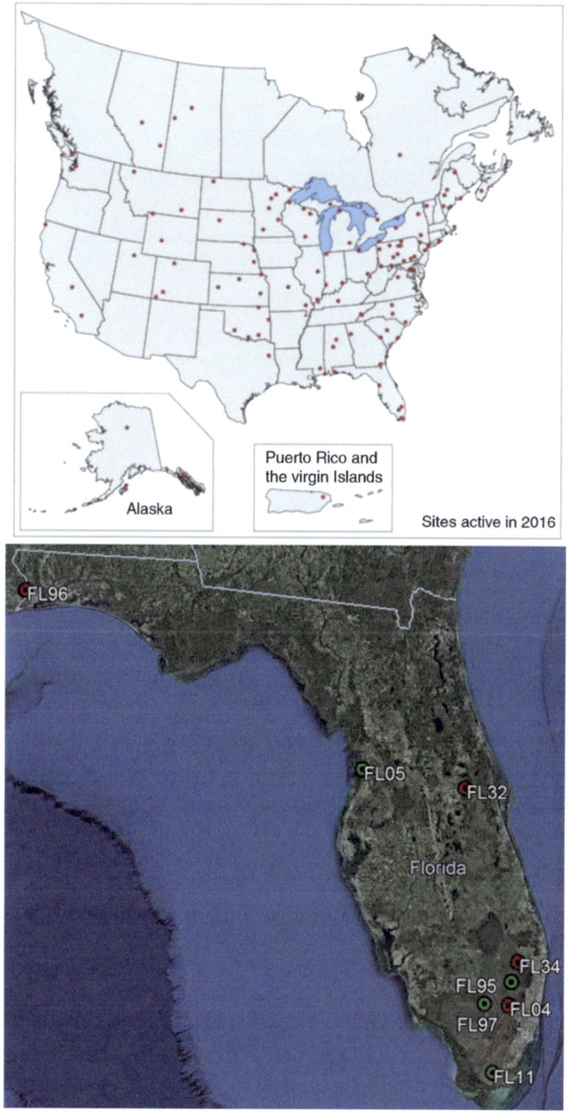

Fig. 4.1 Locations of active MDN Hg wet deposition monitoring stations in 2016 (top) and of all MDN stations historically operated in Florida (bottom) (in bottom panel, green circles = active stations and red circles = inactive stations). (From NADP 2017)

wet deposition (ranging from 17.6 to 25.8 μg/m^2-yr) than most parts of the U.S. with deposition more than double that observed in the western and northeastern U.S. This spatial pattern has been observed for more than 20 years across the MDN network (e.g., Prestbo and Gay 2009). In particular, for the past 20 years, Hg wet deposition has been high at MDN stations in southern Florida with values regularly near or exceeding 18 μg/m^2-yr (Fig. 4.3). In general, some of the highest wet deposition in the country occurs along the Gulf of Mexico; this phenomenon has been attributed to a combination of regional emissions (FDEP 2003) as well as long-

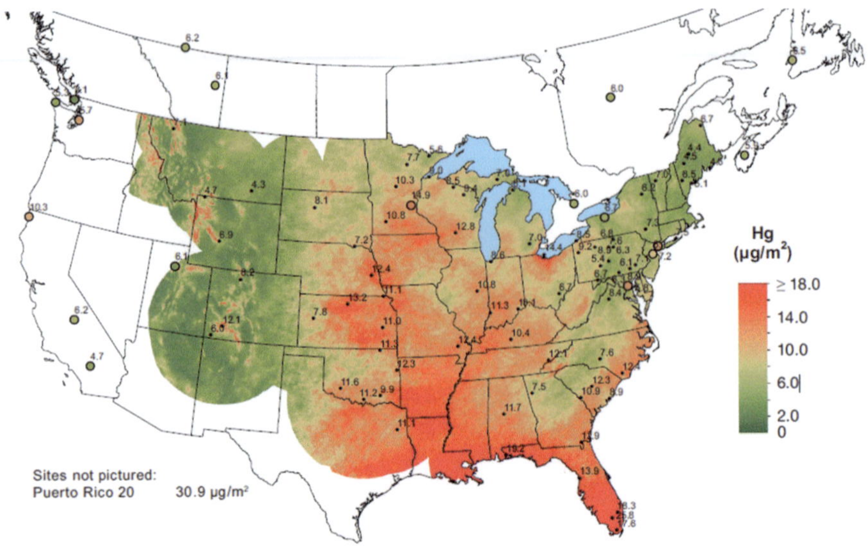

Fig. 4.2 Annual measured Hg wet deposition flux in 2016. (From NADP 2017)

range transport of Hg, high thunderstorm frequency and total rainfall, along with high atmospheric convection during thunderstorms and scavenging of GOM from the free troposphere where high atmospheric concentrations of GOM are evident (FDEP 2013; Nair et al. 2013; Holmes et al. 2016), as well as subsidence of Hg containing air from the free troposphere (Gustin et al. 2012; Holmes et al. 2016; Huang et al. 2017). The high frequency of convective storms (thunderstorms) and the large size of these weather systems in south Florida (storms that can climb in excess of 16 km) allows stripping of atmospheric constituents, including GOM, from these great vertical columns. The idea that convective storm scavenging of high concentrations of GOM in the free troposphere could at least in part be responsible for the fact that wet deposition in south Florida was both elevated and showed no significant east-west gradient moving downwind from the urban fringe along the eastern Florida coast was first proposed by Guentzel et al. (2001). Guentzel et al. predicted that it would only take about 50 pg/m^3 of GOM in the free troposphere upwind from southern Florida to account for the rainfall Hg deposition they observed across southern Florida. This hypothesis subsequently was validated by measurements of GOM obtained by aircraft sampling over the Bahamas which found up to 350 pg/m^3, more than enough to account for the rainfall Hg across the Everglades (Landis et al. 2005).

The prevalence of Hg wet deposition in the Everglades and elsewhere in southern Florida has been investigated in several studies. The Florida Hg Pilot Total Maximum Daily Load (TMDL) study (FDEP 2003) was conducted to evaluate the technical feasibility of calculating an atmospherically driven TMDL for Hg for the Florida Everglades, and the analysis was conducted for a site within the Water Conservation Area 3A (site 3A-15) portion of the Everglades considered to be a "hot-spot" with

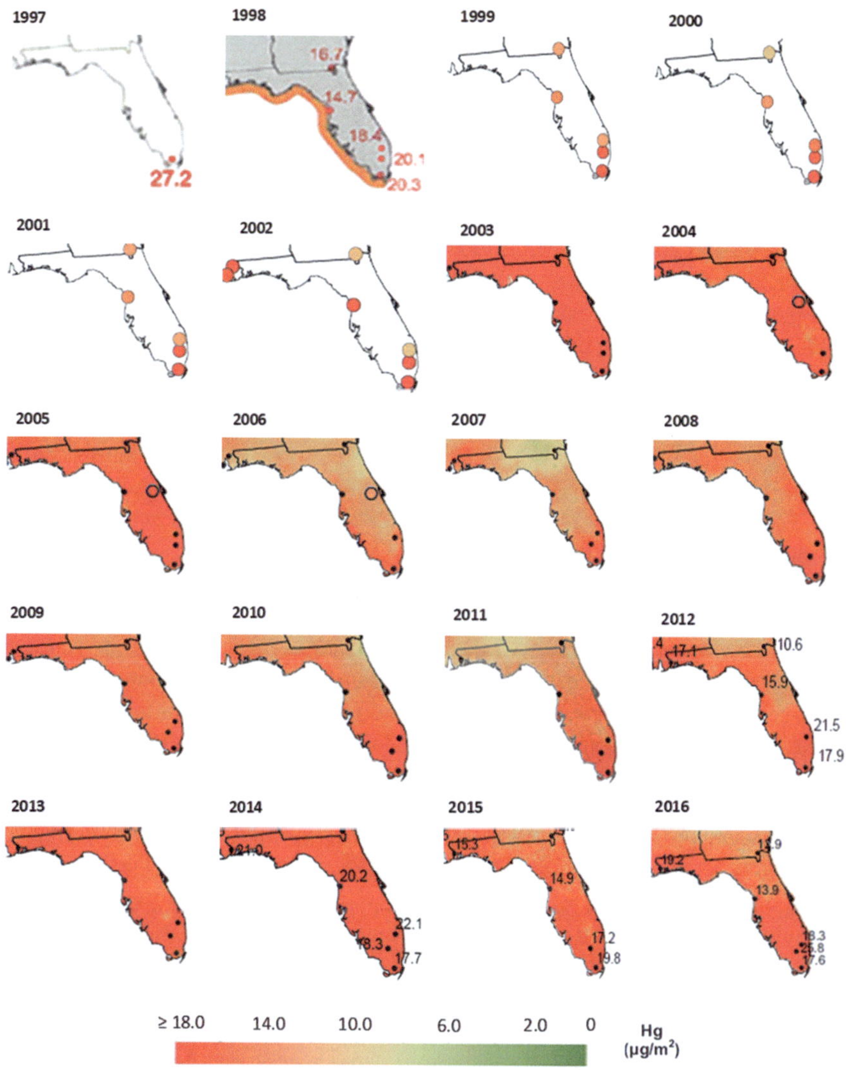

Fig. 4.3 Annual Hg wet deposition flux in Florida from 1997–2016. (From NADP 2017)

respect to fish Hg concentrations. The study incorporated extensive field data into a framework combining atmospheric mercury deposition and aquatic mercury cycling models to demonstrate the feasibility of the approach. The goal was to simulate how changes in local atmospheric mercury emissions in south Florida would influence mercury concentrations in top predator fish, thus demonstrating the potential of combining air and water modeling approaches in TMDLs involving contaminants with an important atmospheric cycling and deposition component such as Hg.

Estimates of atmospheric loading rates of Hg via wet deposition in the Pilot TMDL were obtained from three sources:

1. Measurements of wet deposition at three sites in the Everglades region obtained during the Florida Atmospheric Mercury Study (FAMS) monitoring program conducted between 1993 and 1996 (Guentzel et al. 1995, 2001; Landing et al. 1995; Pollman et al. 1995).
2. Measurements of wet deposition at three MDN sites in south Florida between late 1995 and 2003 (sites FL11, FL04 (Andytown) and FL97 shown in Fig. 4.1).
3. Modeled estimates of wet deposition derived from source-receptor modeling conducted by the University of Michigan Air Quality Laboratory (UMAQL) for the FAMS monitoring period.

During the FAMS program, measurements were made for bulk and wet Hg deposition, particulate-associated mercury and total gas-phase mercury (TGM) at four sites near the Everglades and five other sites in Florida (the panhandle, north-central Florida, a marine background site and two sites in the southwestern peninsula). The FAMS sites used for analysis in the Pilot TMDL were those at Tamiami Trail Ranger Station and Andytown in Florida.

FDEP (2003) combined the Hg wet deposition fluxes, rainfall depth and rain concentrations from the MDN monitor at FL11 from 1996 to 2002 and the equivalent monitor in the FAMS study during 1993–1996 to form a period of record of eight full years. By definition, wet deposition flux is very closely related to rainfall depth and most of the variance in the deposition flux is related to the variance in rainfall depth. Nonetheless, an analysis of volume weighted mean (VWM) concentrations in precipitation revealed that these concentrations declined by approximately 25% (3 ng/L) between 1994 and 2002 due to factors other than seasonal dynamics and precipitation. FDEP (2003) concluded that the relatively modest decreases in VWM concentrations agrees reasonably well with the Hg emissions declines in Dade, Broward and Palm Beach counties from 1993 to 2000.

Axelrad et al. (2005, 2007) subsequently examined whether there had been a continuing decline in atmospheric deposition of mercury in the Everglades beyond 2002 by extending the period of record through 2004. Their analysis showed that an increase in mercury wet deposition and annual VWM concentrations from early 2003 through mid-2004 essentially negated the overall declines that had been observed previously from late 1993 through 2002. Axelrad et al. (2007) also concluded that there was no statistically significant trend in atmospheric deposition during 1994–2005 in south Florida. In a study of regional trends in Hg wet deposition at MDN stations from 1998–2005, Butler et al. (2008) concluded that there was no statistically significant trend in Hg wet deposition or VWM concentration at the FL11 Everglades monitoring station during this period. Similarly, using the non-parametric seasonal Kendall trend test, Prestbo and Gay (2009) determined that there was no statistically significant trend in Hg wet deposition or VWM concentration between 1996 and 2005 at the two stations near the Everglades (FL11 and

FL04). Chapter 1 of Volume III re-examines the issue of recent trends in wet deposition in the Everglades between November 1993 through June 2016 based on FAMS data for FL11 coupled with MDN data. That analysis indicates that the Hg signal declined from late 1993 through mid-2013 followed by a relatively sharp increase through mid-2016.

One feature of the weekly mercury deposition record at the MDN sites near the Everglades is the occurrence of individual weeks with unusually high Hg deposition (Prestbo and Gay 2009). Among all MDN sites in the country, the Everglades station at FL11 shows the highest percent (7%) of its weeks with heavy deposition (with heavy deposition defined here as having annualized wet deposition fluxes greater than 1.5 μg/m^2-yr), reflecting the strong seasonal convective precipitation in this region.

As part of the state-wide Florida mercury TMDL study conducted in 2009 and 2010 (FDEP 2013), the UMAQL measured Hg wet deposition at six monitoring sites across Florida (Fig. 4.4) including one at the ENP at the Beard Research Center (lat. 25.3898, long. -80.6803). Operational logistics of the station were coordinated with ENP staff, with precipitation monitoring beginning in November 2008 and continuing through the end of August 2010. Strong seasonal variability was observed in VWM concentrations as well as Hg wet deposition with increases during summer months (Fig. 4.5), with the sharply accentuated summertime depositional fluxes reflecting higher precipitation volumes coupled with higher VWM concentrations. The annual Hg wet deposition measured at ENP in 2009 was 25 μg/m^2-yr, higher by 2–70% than the other five TMDL sites in Florida.

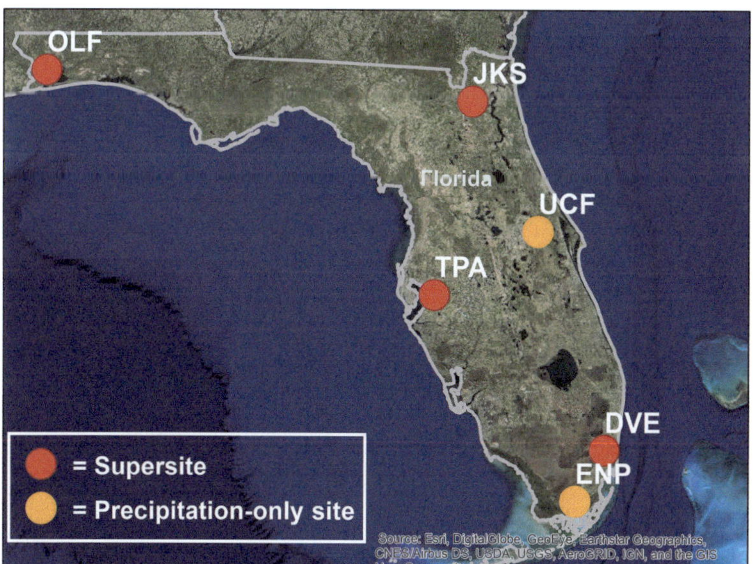

Fig. 4.4 Locations of Hg wet deposition monitoring stations during the 2009 Florida mercury TMDL study. (From FDEP 2013)

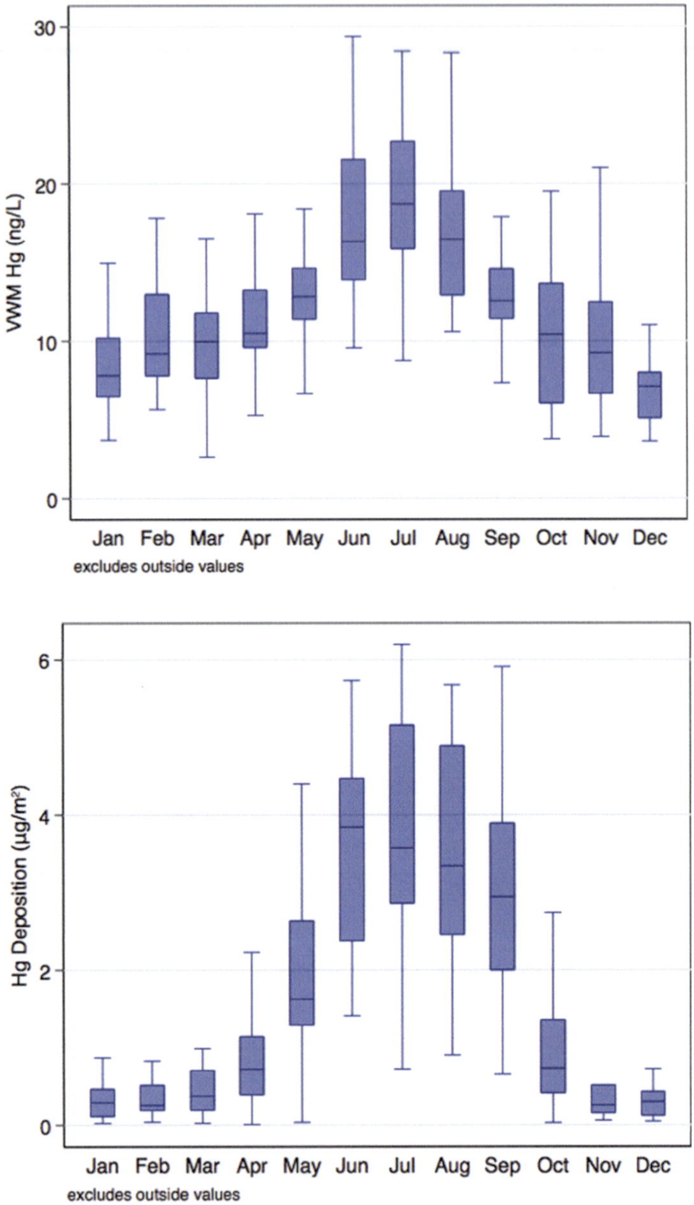

Fig. 4.5 Box plots of the distribution of monthly volume weighted mean Hg concentration in wet deposition (top) and Hg wet deposition flux (bottom) at the Everglades (ENP) monitoring site, November 1993 through June 2016. (Data from FAMS (Guentzel et al. 2001; NADP 2017))

The discussions of Hg wet deposition above have been for inorganic Hg. Measurements of MeHg deposition are much more scant than those of inorganic Hg. MeHg is measured at a few MDN stations none of which are in Florida. The concentration of MeHg in summertime precipitation measured at south Florida sites during FAMS ranged from <0.005 to 0.020 ng/L based on limited wet and bulk deposition measurements conducted during the FAMS study (Guentzel et al. 1995). These values were 2–3 orders of magnitude lower than corresponding inorganic Hg wet deposition. The FAMS study concluded that deposition of MeHg may not be a significant source of MeHg in south Florida's aquatic environments.

4.3 Dry Deposition

While wet deposition can be well quantified through the collection and analysis of precipitation, direct measurements of dry deposition are difficult to make and have large uncertainties (e.g., Lyman et al. 2007; Zhang et al. 2009, 2012; Lai et al. 2011). FDEP (2013) provide a succinct explanation of the importance of dry deposition measurements in Florida that are applicable to the Everglades. "The dry deposition mercury speciation and continuous measures are important in understanding the specifics and dynamics of mercury cycling within Florida. Atmospheric dry mercury is stripped by forests in leaf and needle uptake as well as in resistance knocking *(removing)* mercury from the air to the forest floor". The atmospheric dry deposition of Hg to prairie, shrub, and wetland plants may be a critical avenue of entry into food webs, and a potential first step of Hg being bound to organic matter and eventually entering aquatic systems.

Although measurements of Hg dry deposition are scarce to this day and measurements of wet deposition were uncommon in the Everglades prior to the 1990s, cores of lake sediments and bog peat have been used to document historical changes in atmospheric Hg deposition, both wet and dry. Analyses of sediment records in the Everglades indicate an increasing trend in total deposition up through the decade of the 1980s (Rood et al. 1995; Delfino et al. 1993; see Chap. 1, Vol. III for a review of these studies). Pollman et al. (2007) inferred historic atmospheric fluxes of Hg over a 200-year period from 1800–2000 using sediment cores collected at site F3 in Water Conservation Area (WCA-2A) (Fig. 4.6) using unpublished data from Abelak et al. provided through T. Atkeson (FDEP). There is a clear increasing trend in total deposition over the 200-year period till the 1980s followed by a reduction in total deposition.

A modeled analysis of dry deposition conducted during the Florida Pilot TMDL (FDEP 2003) suggests that dry deposition is an important component, comprising perhaps 34–40% of the total Hg deposition signal. Dry deposition to the Everglades

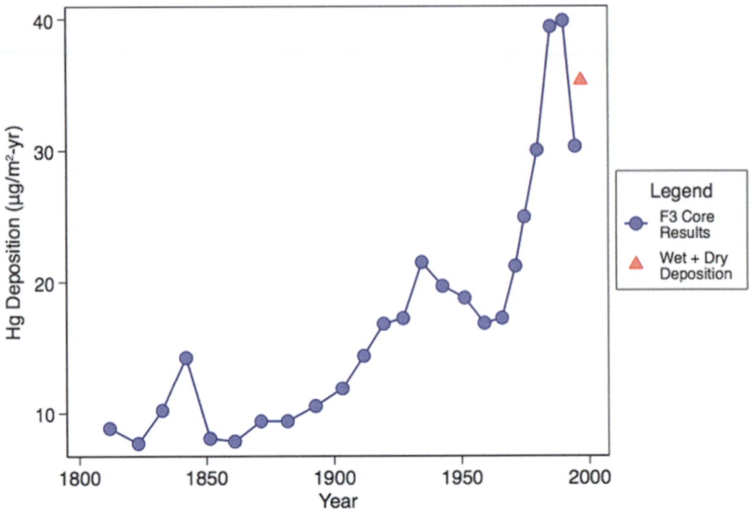

Fig. 4.6 Inferred atmospheric fluxes of Hg from sediment core collected at site F3 in WCA-2A. Red triangle shows for comparison the estimated total deposition flux (wet and dry) of Hg based on FAMS measurements of wet deposition and dry deposition fluxes modeled by Keeler et al. (2000). (Data from Abelak et al. (unpublished data). Redrawn from Pollman et al. (2007))

WCA 3A-15 was estimated using a hybrid model system comprising the Hybrid Single Particle Lagrangian Integrated Trajectories Model, version 4 (HYSPLIT_4) and the Regional Atmospheric Modeling System (RAMS) (FDEP 2003). Dry deposition due to local sources was estimated to be 12.2 ± 2.0 µg/m^2-yr (+1 standard deviation) during a 12-month period from June 1995 to June 1996. While considerable variability was evident in the monthly deposition estimates, on average, dry deposition to the site showed a seasonal trend, with relatively greater deposition occurring during the climatological wet season. As was the case for the wet deposition to WCA 3A-15, dry deposition to this area was dominated by the GOM fraction (FDEP 2003).

Atkeson et al. (2005) combined dry deposition estimates modeled by Keeler et al. (2000) for 22 June 1995 through 21 June 1996 with wet deposition measured at FAMS sites during the same period to obtain a total deposition flux of 35.3 µg/m^2-yr. Pollman et al. (2007) estimated that the dry deposition flux in 2000 at ENP was 10.1 µg/m^2-yr using the measured wet deposition flux of 19.9 µg/m^2-yr at the Beard Research Center (FL11) MDN site in ENP and assuming that the dry:wet deposition ratio of 0.528 used by Atkeson et al. (2005) had remained constant.

During the Florida statewide mercury TMDL study (FDEP 2013), an inferential model was used to estimate Hg dry deposition (Marsik et al. 2011) using surface measurements of ambient speciated Hg and meteorological variables at four “supersites” (Pensacola, Jacksonville, Tampa, and Davie) for 14–18 months from 2009 to

2010. The annual dry deposition flux at Davie was estimated to be 31 μg/m^2-yr, lower than the values in central and northern Florida (FDEP 2013), but higher than the value estimated above for the ENP.

4.4 Summary

The robust long-term record of atmospheric wet deposition of Hg to the Everglades from the NADP Mercury Deposition Network indicates that inorganic Hg wet deposition in this region has continued to be among the highest in the country over the past two decades. In particular, measurements at the Beard Research Center at the Everglades National Park show annual wet deposition exceeding 18 μg/m^2-yr (and often higher than 20 μg/m^2-yr) during most of the time since monitoring began in 1996. These high values have also been confirmed in periodic monitoring campaigns such as the Florida Pilot and State-wide Hg TMDL programs. In contrast, the Hg wet deposition flux over most of the rest of the country is much lower and varies between 5 and 15 μg/m^2-yr. The strong summertime convective precipitation in the Everglades and high concentrations of Hg in precipitation result in individual weeks with unusually high mercury deposition, a characteristic feature of wet deposition dynamics in the Everglades.

Observations of Hg dry deposition are scarce due to challenges in quantifying Hg molecular diffusion and removal; however, the limited measurements (and associated modeling) suggest that the dry deposition signal to the Everglades could be comparable to that of wet deposition. The significantly different deposition characteristics of the different forms of Hg in the atmosphere and the differences in their subsequent methylation characteristics underpin the need to understand not just the magnitude of total deposition to the Everglades but the forms of Hg present in deposition. The dry deposition measurements of Hg are likely an underestimate as they do not typically include dry deposition of GEM that has the potential to be later converted to GOM. Although the deposition of MeHg to the Everglades is likely two orders of magnitude lower than that of inorganic Hg, the organic form needs to be studied as well because it has the potential to directly affect the Everglades without a need for methylation first.

Declines in the wet deposition and Hg concentrations in precipitation in the Everglades in the 1990s appear to be correlated with declines in local Hg emission sources. These trends in Hg deposition are investigated further in Chap. 1, Volume III. The continued high wet deposition to south Florida in the 2000s and 2010s despite significant changes in local and other Florida Hg emissions suggest a strong long-range contribution to Hg deposition in the Everglades. This concept is explored further in Chap. 5 (this volume).

References

Atkeson TD, Pollman CD Axelrad DR (2005) Chapter 26. Recent trends in mercury emissions, deposition, and biota in the Florida Everglades: a monitoring and modeling analysis. In: Pirrone N and Mahaffey KR [eds] Dynamics of mercury pollution on regional and global scales: atmospheric processes and human exposures around the world, Springer, New York, pp 637–655

Axelrad DM, Atkeson TD, Pollman CD, Lange T, Rumbold DG, Weaver K (2005) Chapter 2B: mercury monitoring, research and environmental assessment in South Florida. In: Redfield G (ed) 2005 South Florida Environmental Report. South Florida Water Management District, West Palm Beach

Axelrad DM, Atkeson TD, Lange T, Pollman CD, Gilmour CC, Orem WH, Mendelssohn IA, Frederick PC, Krabbenhoft DP, Aiken GR, Rumbold DG, Scheidt DJ, Kalla PI (2007) South Florida Environmental Report. Chapter 3B: mercury monitoring, research and environmental assessment in South Florida

Butler TJ, Cohen MD, Vermeylen FM, Likens GE, Schmeltz D, Artz RS (2008) Regional precipitation mercury trends in the eastern USA, 1998–2005: declines in the Northeast and Midwest, no trend in the Southeast. Atmos Environ 42:1582–1592

Delfino JJ, Crisman TL, Gottgens JF, Rood BR, Earle CDA (1993) Spatial and temporal distribution of mercury in Everglades and Okefenokee wetland sediments. Final Project Report (April 1, 1991 - June 30, 1993) to South Florida Water Management District (contract no. C91–2237), USGS (contract no. 14-08-0001-G-2012) and Florida DER (contract no. WM415)

Florida Department of Environmental Protection (FDEP) (2003) Integrating atmospheric mercury deposition and aquatic cycling in the Florida Everglades: an approach for conducting a Total maximum daily load analysis for an atmospherically derived pollutant. Final report. Authors: Atkeson T, Axelrad D, Pollman CD, Keeler G. Prepared for United States Environmental Protection Agency. October 2003

Florida Department of Environmental Protection (FDEP) (2013) Mercury TMDL for the State of Florida. Final report. Watershed evaluation and TMDL Section. October 24. https://floridadep.gov/sites/default/files/Mercury-TMDL.pdf

Guentzel JL, Landing WM, Gill GA, Pollman CD (1995) Atmospheric deposition of mercury in Florida: the FAMS project (1992–1994). Water Air Soil Pollut 80:393–402

Guentzel JL, Landing WM, Gill GA, Pollman CD (1998) Mercury and major ions in rainfall, throughfall, and foliage from the Florida Everglades. Sci Total Environ 213:43–51

Guentzel JL, Landing WM, Gill GA, Pollman CD (2001) Processes influencing rainfall deposition of mercury in Florida. Environ Sci Technol 35:863–873

Gustin MS, Weiss-Penzias PS, Peterson C (2012) Investigating sources of gaseous oxidized mercury in dry deposition at three sites across Florida, USA. Atmos Chem Phys 12:9201–9219

Holmes CD, Krishnamurthy NP, Caffrey JM, Landing WM, Edgerton ES, Knapp KR, Nair US (2016) Thunderstorms increase mercury wet deposition. Environ Sci Technol 50(17):9343–9350. https://doi.org/10.1021/acs.est.6b02586

Huang J, Miller MB, Edgerton E, Gustin MS (2017) Deciphering potential chemical compounds of gaseous oxidized mercury in Florida, USA. Atmos Chem Phys 17:1689–1698

Keeler GJ, Marsik FJ, Al-Wali KI, Dvonch JT (2000) Project final technical report – modeled deposition of Speciated mercury to the SFWMD water conservation area 3A: 22 June 1995 to 21 June 1996. Project description and results. The University of Michigan Air Quality Laboratory. Ann Arbor, Michigan

Lai S-O, Huang J, Holsen TM, Hopke PK (2011) An evaluation of direct measurement techniques for mercury dry deposition. Sci Total Environ 409:1320–1327

Landing W, Perry JJ, Guentzel JL, Gill GA, Pollman CD (1995) Relationships between the atmospheric deposition of trace elements, major ions, and mercury in Florida: the FAMS project (1992–1993). Water Air Soil Pollut 80:343–352

Landis MS, Lynam MM, Stevens RK (2005) Chapter 7. The monitoring and modelling of mercury species in support of local, regional and global modelling. In: Pirrone N, Mahaffey KR

(eds) Dynamics of mercury pollution on regional and global scales: atmospheric processes and human exposures around the world. Springer, New York, pp 1233–1151
Lyman SN, Gustin MS, Prestbo EM, Marsik FJ (2007) Estimation of dry deposition of atmospheric mercury in Nevada by direct and indirect methods. Environ Sci Technol 41:1970–1976
Marsik FJ, Keeler GJ, Dvonch JT, Sillman S, Pollman C, Pirrone N, Hedgecock I, Jung G, Schürmann G (2011) Monthly inferential modeled estimates of mercury dry deposition. Report to Florida Department of Environmental Protection. pp 1–12
Nair US, Wu Y, Holmes CD, Ter Schure A, Kallos G, Walters JT (2013) Cloud-resolving simulations of mercury scavenging and deposition in thunderstorms. Atmos Chem Phys 13:10143–10157
National Atmospheric Deposition Program (NADP) (2017) Mercury Deposition Network (MDN) Data available at http://nadp.sws.uiuc.edu/mdn. NADP Program Office. Illinois State Water Survey, University of Illinois, Champaign
Pollman CD, Gill GA, Landing Guentzel WM, Bare D, Porcella D, Zillioux E, Atkeson T (1995) Overview of the Florida Atmospheric Mercury Study (FAMS). Water Air Soil Pollut 80:285–290
Pollman CD, Porcella DM, Engstrom DR (2007) Assessment of trends in mercury-related data sets and critical assessments of cause and effect for trends in mercury concentrations in Florida biota: phase II. Final report submitted to the Florida electric power coordinating group and the Florida Department of Environmental Protection. 17 July. p 59
Prestbo E, Gay D (2009) Wet deposition of mercury in the U.S. and Canada, 1996–2005: results and analysis of the NADP mercury deposition network (MDN). Atmos Environ 43:4223–4233
PRISM Climate Group (2017) Oregon State University. http://prism.oregonstate.edu
Rood BE, Gottgens JF, Delfino JJ, Earle CD, Crisman TL (1995) Mercury accumulation trends in Florida Everglades and Savannas marsh flooded soils. Water Air Soil Pollut 80(1–4):981–990
United States Environmental Protection Agency (USEPA) (1996) South Florida ecosystem assessment interim report. Monitoring for adaptive management: implications for ecosystem restoration. Region 4 SESD and ORD. EPA-904-96-008. December 1996
Zhang L, Blanchard P, Johnson D, Dastoor A, Ryzhkov A, Lin CJ, Vijayaraghavan K, Gay D, Holsen TM, Huang J, Graydon JA, Louis VLS, Castro MS, Miller EK, Marsik F, Lu J, Poissant L, Pilote M, Zhang KM (2012) Assessment of modeled mercury dry deposition over the Great Lakes region. Environ Pollut 161:272–283
Zhang L, Wright LP, Blanchard P (2009) A review of current knowledge concerning dry deposition of atmospheric mercury. Atmos Environ 43:5853–5864

Chapter 5
Mercury Emission Sources and Contributions of Atmospheric Deposition to the Everglades

Krish Vijayaraghavan and Curtis D. Pollman

Abstract This chapter presents a summary of the literature on mercury (Hg) air emission sources in south Florida, the rest of the state, and the rest of the United States and the world, and their historical and current contributions to atmospheric deposition of Hg in the Everglades from monitoring and computer modeling. Hg deposited to the Everglades is a combination of gaseous oxidized Hg and particulate Hg emitted locally and gaseous elemental Hg emitted by distant sources and transported to south Florida where it is converted to the oxidized form that deposits more readily. While local emission sources of Hg likely contributed a large fraction (30% to over 45%) of deposition to the south Florida peninsula till the 1990s, stringent control measures have since been implemented on these sources and Hg reduced in commonly used materials. Currently, less than 1% of Hg deposition to the Everglades is likely due to Florida sources, with 85–95% due to the long-range transport of Hg from non-US sources. Summertime thunderstorms represent a key mode for the transfer of the Hg transported from distant sources to the Everglades as they reach into the upper troposphere where the oxidized form of Hg resides and mix it down to lower altitudes from where it is deposited to the Everglades under wet and dry conditions.

Keywords Source attribution · Modeling · Convective storms · Thunderstorms · Long-range transport · Atmospheric mercury transformations

K. Vijayaraghavan (✉)
Ramboll US Corporation, Novato, CA, USA
e-mail: kvijay@ramboll.com

C. D. Pollman
Aqua Lux Lucis, Inc, Gainesville, FL, USA
e-mail: cpollman@aqualuxlucis.org

C. D. Pollman et al. (eds.), *Mercury and the Everglades. A Synthesis and Model for Complex Ecosystem Restoration*,
https://doi.org/10.1007/978-3-030-20070-1_5

5.1 Introduction

As the deposition of inorganic mercury (Hg) emitted to the atmosphere represents most of the external contribution of Hg to the Everglades, it is important to understand the potential sources of these emissions. Hg emissions to the atmosphere result from the release during anthropogenic activities or natural phenomena of Hg present in minerals and soil, vegetation and water, and due to the release of Hg in products and industrial processes. The speciation of Hg emissions varies by source category and affects the eventual fate of these releases (see Chap. 3, this volume). Due to its long half-life in the atmosphere (approximately 6–12 months), gaseous elemental Hg [GEM or Hg(0)] emitted from sources far from the Everglades (such as in other US states and other countries) undergoes long-range transport, and can contribute to atmospheric deposition in the Everglades. Gaseous oxidized Hg [GOM or Hg(II)] emissions exhibit more local-scale deposition while the deposition of particulate bound Hg [PBM or Hg(p)] depends on the size of the particles to which Hg is bound, with coarse PBM deposition occurring primarily due to local sources. The elemental and divalent forms of Hg convert between each other, thus further affecting these deposition rates. Thus, when identifying the emission sources contributing to Hg contamination in the Everglades, it is important to examine not only local sources but also other sources in Florida, elsewhere in the US and the rest of the world. The speciation of Hg emissions also may change with the application of control technologies due to chemical transformation. In particular, coal-fired electric utilities could have very different Hg speciation depending on coal quality and controls implemented. For example, the use of a selective catalytic reduction (SCR) system increases the fraction of GOM due to oxidation from GEM while the use of a scrubber decreases the fraction of GOM and PBM. Because the effect of Hg emissions on ecosystems such as the Everglades may be seen for decades past their time of release (see Chap. 3, Volume III for a discussion of both legacy Hg deposits and the importance of newly deposited Hg in aquatic ecosystems), it is critical to consider the contributions of not only sources currently in operation, but also historical sources of Hg emissions in the Everglades region and further afield as discussed below.

5.2 Mercury Sources and Emissions in South Florida

An emissions inventory of Hg point and non-point sources for 16 counties in south Florida for 1996 (RMB 1999) identified soil degassing as the largest emission source category for both anthropogenic and natural sources (7158 kg/yr) while municipal waste combustors (MWC) were the largest anthropogenic contributor at (4445 kg/yr or 82% of all anthropogenic emissions) followed by power generation facilities, medical waste incinerators (MWI), and sugar refineries (435,213 and 136 kg/yr, respectively). Emissions from Broward, Miami-Dade and Palm Beach counties comprised nearly half (49%) of the total Hg emissions in the 16-county area. Emissions from these three counties were identified as important contributors to Hg deposition in the Everglades through a multivariate receptor modeling analysis of wet deposition measurements (Dvonch et al. 1999). Therefore, a subsequent emissions inventory study in south Florida (RMB 2002) focused on Hg emissions in these three counties and across a 20-year historical period from 1980 to 2000. This historical inventory included the three largest emission source groups (MWC, MWI and sugar refineries) in the 3-county area as well as power generation facilities that were negligible in this area but were of interest for many observers.

Hg emissions in the three counties with large historical contributions to Hg deposition in the Everglades increased three-fold from 1982 to 1983 as MWCs and MWIs became operational (Fig. 5.1). Emissions from these two source categories continued to increase through 1991 when total Hg emissions in the 3-county area were estimated to have reached their peak of 3077 kg/yr. After the USEPA promulgated new source performance standards (NSPS) in mid-1992 and subsequently required installation of Hg control technology, many MWIs ceased operations and medical waste was either autoclaved for steam sterilization or sent offsite for processing (RMB 2002). Actual test data were used to estimate emissions for MWCs and the reductions in emissions in the mid- to late-1990s are believed to be due to the implementation of controls such as carbon injection (RMB 2002). Total emissions from MWIs and MWCs declined by 93% from 1991 to 2000. Over the entire 20-year period examined, MWIs and MWCs together represented over half of the total emissions and as much as 96% of the total in 1990–1992. During the relatively high (> 1000 kg/yr) emission years of 1983–1994, sugar processing and power generation typically constituted less than 10% and 1%, respectively, of the total.

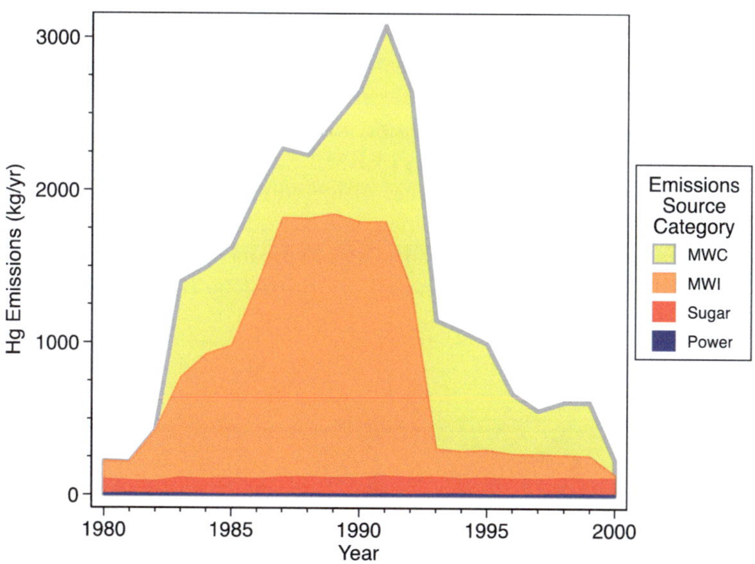

Fig. 5.1 Estimated annual Hg emissions during 1980–2000 from power generation, sugar processing facilities, medical waste incinerators and municipal waste combustors in Broward, Miami-Dade and Palm Beach counties. (Data source: RMB 2002)

While the inventory described above did not include Hg speciation, Dvonch et al. (1999) estimated the GEM/GOM/PBM split to be approximately 4/95/1% for MWIs and 24/75/1% for MWCs. Because emissions from MWIs and MWCs have such a large fraction of GOM (the form of Hg that tends to deposit near sources), these two source groups in the 3-county area likely had a large historical contribution to the Everglades during their period of operation.

A different approach to constructing a Hg emissions inventory was adopted by Husar and Husar (2002) who estimated the historical annual amount of Hg mobilized in fuels and consumer and commercial products, and inferred emissions to the atmosphere using information on fuel consumption and Hg use/content in the production of Hg containing products. These estimates were made for the US and then applied to the regional level in Florida for 1930–2000. Similar to the RMB (2002) study, Hg chemical speciation was not estimated. In addition, Husar and Husar (2002) did not consider the agricultural sector. Water-based paint applications dominated Hg-containing products from the 1960s to 1990 for Broward County and Miami-Dade County (and elsewhere in Florida; Fig. 5.2) and then abruptly declined when USEPA banned Hg use in paint in 1990. Husar and Husar (2002) assumed that 75% of the Hg in paint used was released to the atmosphere in the first year of use. They also assumed that Hg in products categories (electrical, controls, laboratory) were disposed as municipal solid waste (MSW). The fraction of waste burned was based on Florida Department of Environmental Protection (FDEP) reports for 1990–1998 and assumed to range from 10% to 30% for earlier years. The second largest Hg contributions to Broward and Miami-Dade were associated with the disposal of electrical (battery) devices.

The estimates for coal contribution to Hg emissions for Broward, Miami-Dade and Palm Beach counties were assumed to be negligible and were not included. Using the assumption that electrical, control, and laboratory use category will end up in MSW, Husar and Husar (2002) estimated the sum of Hg emissions for Broward, Dade, and Palm Beach Counties and compared these with the MWI and MWC (i.e., MSW) emission estimates of RMB (2002) (Fig. 5.3). Here, MSW_R represents the municipal solid waste incineration from RMB (2002) while MSW_R + MWI_R is the sum including medical waste incineration. The RMB Hg emission estimates are approximately 10–50% of the mercury amount mobilized in electrical use, laboratory use, and controls as estimated by Husar and Husar (2002)

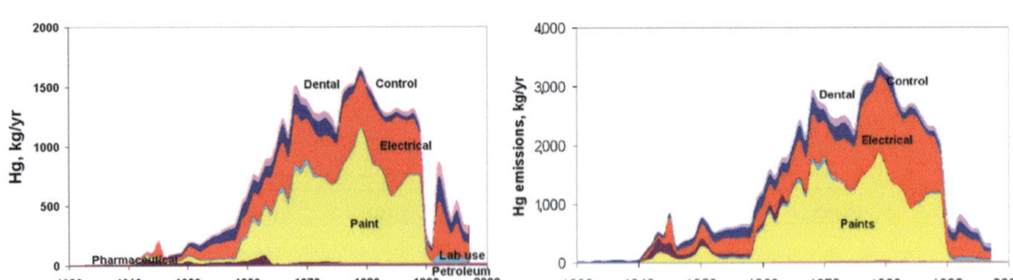

Fig. 5.2 Annual Hg releases during 1930–2000 estimated from mercury mobilized in fuels and goods in Broward (left) and Miami-Dade (right) counties. (From Husar and Husar 2002)

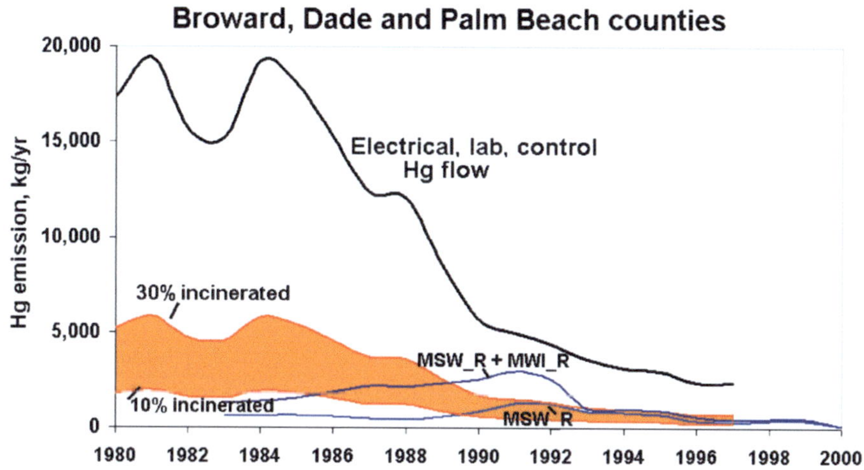

Fig. 5.3 Sum of municipal solid waste and medical waste incineration emissions for Broward, Miami-Dade, and Palm Beach counties. (From Husar and Husar 2002)

because only a fraction of the Hg in goods is incinerated and emitted. The shaded area in Fig. 5.3 represents a crude emission estimate by Husar and Husar (2002) assuming that 15–30% of Hg in goods is incinerated; the emissions estimates at the lower end of that range are generally comparable with those of RMB (2002). Husar and Husar (2002) judged the estimation of county Hg emissions from the Hg mass mobilization data to be the most uncertain part of their entire analysis.

5.2.1 Emissions Trends for South Florida from the USEPA National Emissions Inventory

The USEPA National Emissions Inventory (NEI) provides a detailed estimate of air emissions of criteria pollutants and precursors, and hazardous air pollutants including mercury.[1] The NEI is released every 3 years based primarily upon data provided by state, local, and tribal air agencies for sources in their jurisdictions and supplemented by data developed by the USEPA. Figure 5.4 and Table 5.1 present a summary of facilities with mercury emissions exceeding 10 lb/yr from the 2002, 2005, 2008, 2011 and 2014 NEI data for southern Florida counties with Everglades habitat. A reduction in the number of such facilities is evident over this 13-year period largely reflecting increasing controls on Hg emissions and closure of units with large sources mainly in Miami-Dade, Broward and Hendry counties. When considering emissions from all facilities (i.e., those at all emissions levels) in this region, emissions declined by 53% from 828 lb/yr in 2002 to 391 lb/yr in 2011 and have then subsequently increased by 6% in 2014 (Table 5.2). While emissions from

[1] https://www.epa.gov/air-emissions-inventories/national-emissions-inventory-nei

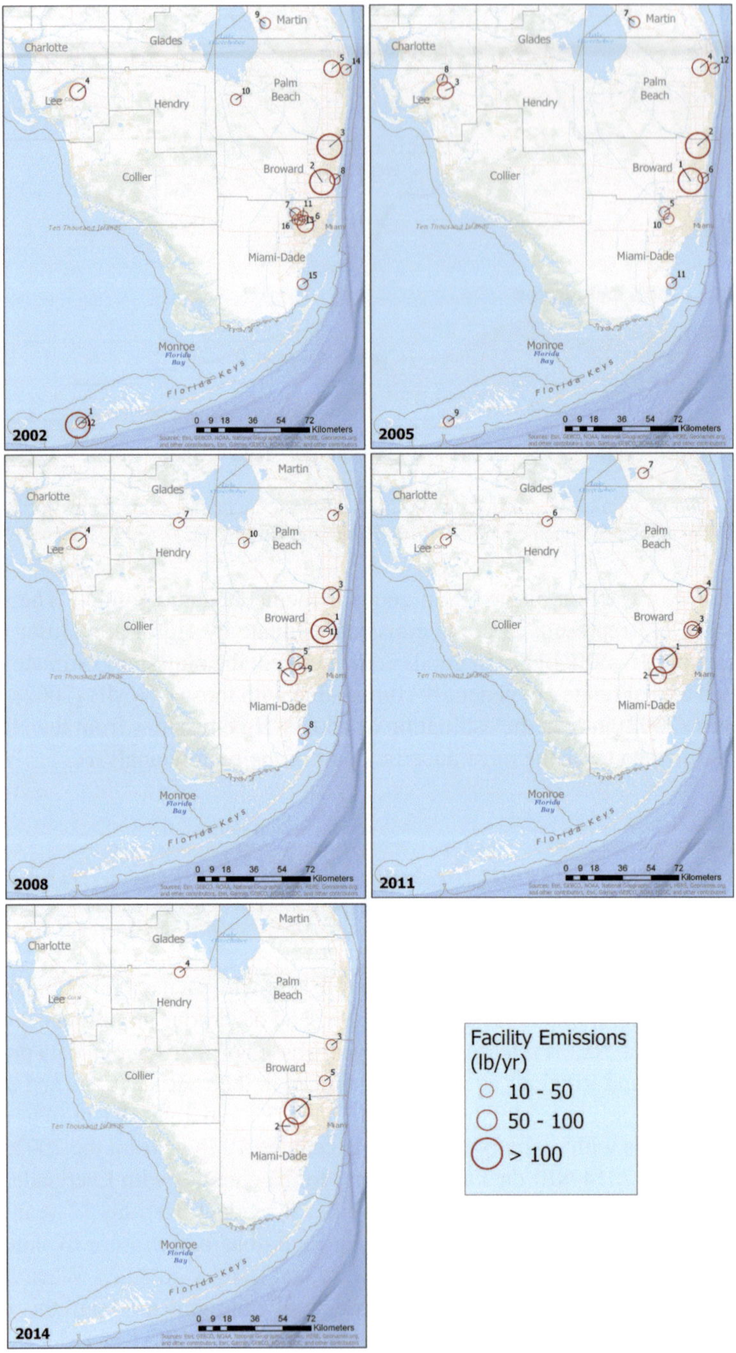

Fig. 5.4 Facilities with mercury emissions exceeding 10 lb/yr from National Emissions Inventory data for 2002, 2005, 2008, 2011, and 2014 for southern Florida counties (see Table 5.1 for emission values)

Table 5.1 Facilities with Hg emissions exceeding 10 lb/yr from NEI data for southern Florida counties (see Fig. 5.4 for locations) (note: ranks in different years correspond to different facilities)[a]

Rank	2002 Facility type	2002 Hg (lb/yr)	2005 Facility type	2005 Hg (lb/yr)	2008 Facility type	2008 Hg (lb/yr)	2011 Facility type	2011 Hg (lb/yr)	2014 Facility type	2014 Hg (lb/yr)
1	MWC	151.4	MWC	116.9	MWC	141.8	Cement	112.0	Cement	222.0
2	MWC	116.8	MWC	100.6	Cement	84.0	Cement	61.0	Cement	74.0
3	MWC	100.6	MWC	96.2	MWC	81.7	MWC	55.7	MWC	35.1
4	MWC	96.2	MWC	73.8	MWC	73.0	MWC	54.2	Sugar mill	22.0
5	MWC	73.8	Cement	43.1	Cement	69.8	MWC	26.9	FFEPG	10.4
6	MWC	60.6	FFEPG	28.5	MWC	42.0	Sugar mill	19.0		
7	Cement	44.5	FFEPG	22.4	Sugar mill	20.6	FFEPG	18.0		
8	FFEPG	29.5	FFEPG	18.3	Nuclear	16.0	FFEPG	10.9		
9	FFEPG	24.8	MWC	15.6	MWC	15.5				
10	Biomass	22.9	MWC	15.2	Biomass	13.5				
11	Foundry	19.7	Nuclear	13.1	FFEPG	10.8				
12	MWC	15.6	FFEPG	11.8						
13	MWC	15.2								
14	FFEPG	13.9								
15	Nuclear	13.4								
16	MWC	12.6								

[a]*FFEPG* Fossil fuel electric power generation, *MWC* Municipal waste combustion

Table 5.2 Mercury emissions from facilities in southern Florida (lb/yr) derived from NEI estimates

Year	2002	2005	2008	2011	2014
Hg emissions (lb/yr)	828.3	562.5	592.5	391.0	415.1

MWCs and electric power generation have steadily decreased, cement manufacturing emissions have increased over the past decade and now represent the largest point source contributor to Hg in southern Florida (296 lb/yr in 2014).

5.3 Mercury Sources and Emissions in the Rest of Florida

As in south Florida, Hg sources and emissions in the rest of Florida have changed markedly over the last four decades due to regulations and the implementation of controls. Hg releases from the disposal of electrical devices dominated Hg releases in the state in the late twentieth century (Husar and Husar 2002) (Fig. 5.5) with Hg due to batteries in electrical devices peaking at over 50 Mg/yr. in the late 1980s (Fig. 5.6). The next largest category of Hg use in products was in water soluble

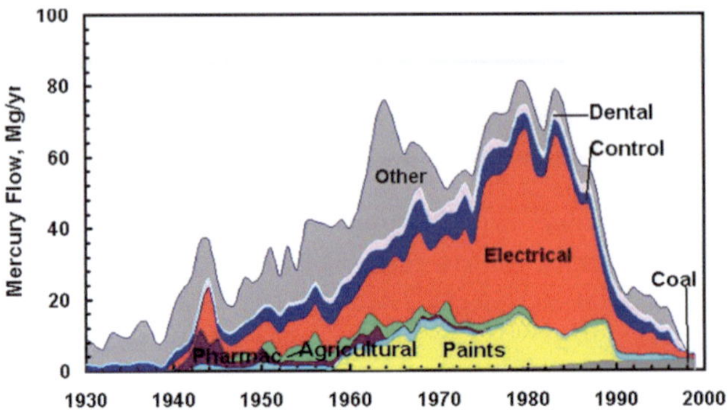

Fig. 5.5 Annual Hg releases inferred from material flow by major use category in Florida during 1930–2000. (From Husar and Husar 2002)

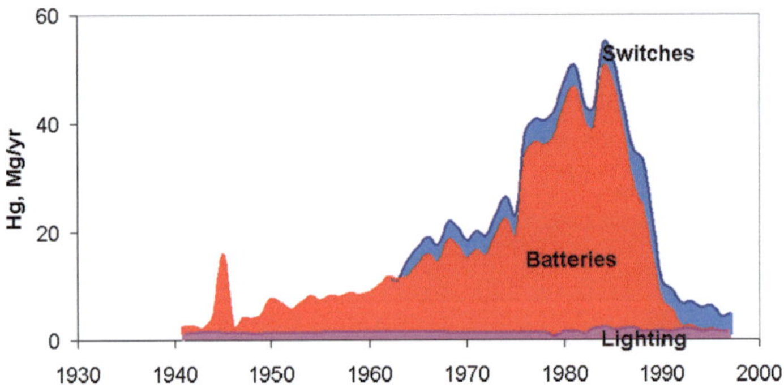

Fig. 5.6 Annual Hg releases from electrical device use in Florida during 1930–2000. (From Husar and Husar 2002)

paints due to the addition of organomercury compounds, and phenylmercuric acetate in particular, as fungicides to prolong the paint's shelf life, resulting in potentially over 10 Mg/yr emissions (Husar and Husar 2002) also in the late 1980s. Since that time, Hg has almost been eliminated from batteries and latex paints (FDEP 2013). Other historical sources of Hg emissions in Florida from electrical device use include the disposal of fluorescent and high intensity discharge (HID) lights and wiring devices and electric light switches. Non-Hg switches were introduced in the 1990s. However, switches have a life expectancy of up to 50 years (Husar and Husar 2002). Therefore, Hg releases due to the disposal of legacy Hg-containing electric light switches in Florida will continue through the 2030s.

Estimates of Hg emissions from coal combustion in Florida by Husar and Husar (2002), based on annual coal production, coal Hg content, the fraction of Hg released and various control efficiencies, show a steady, albeit small, increase from 1980 to 2000; these estimates compare well with measurements of Hg emissions made during the 1999 USEPA Information Collection Request (ICR) at power plants. Depending on the estimated Hg content of coal, the Hg emissions from coal combustion in Florida in 2000 ranged between 1.4 and 1.9 Mg/yr.

Table 5.3 Mercury emissions from major source categories in Florida in 2005, 2009 and 2014

Source category	2005 (from FDEP, 2013)		2009 (from FDEP, 2013)		2014 (from USEPA NEI)	
	lb/yr	% of total	lb/yr	% of total	lb/yr	% of total
Coal-fired electric generation	2094	53%	1469	46%	1185	36%
Cement Industry	710	18%	326	10%	559	17%
Waste to energy plants	692	17%	663	21%	252	8%
Other electric generation	314	8%	314	10%	310	10%
Waste water treatment plants	102	3%	102	3%	25	1%
All others	60	2%	295	9%	916	28%
Total	3972		3169		3247	

Adapted from FDEP, 2013 and the USEPA NEI

More detailed estimates of Hg emissions from coal combustion since 2000 are available from the NEI. Table 5.3 shows the major source categories of anthropogenic Hg emissions in Florida from the 2005 and 2014 NEI databases and from a 2009 estimate from the state-wide Florida Total Maximum Daily Load (TMDL) determination. As noted by FDEP (2013) and seen in Table 5.3, Hg emissions from coal combustion at Florida utilities have declined over the past 10 years or so because lower natural gas prices have spurred many utilities that historically relied on coal to use natural gas (and which has a Hg content that is negligible compared to coal) by either retrofitting existing units to use natural gas, constructing new gas-fired units or operating existing natural gas electricity generating units (EGUs) at higher capacity. In addition to lower natural gas prices, Hg emissions from Florida coal-fired power plants have generally steadily declined due to the installation of controls prior or in anticipation of implementation of the Mercury and Air Toxics Standards (MATS), with reductions between 57% and 90% at several plants (FDEP 2013).

Emissions from waste-to-energy plants have decreased due to the implementation of Hg controls as well. However, the reduction in Hg emissions from cement production from 2005 to 2009 was due to a slowing of the housing market rather than controls (FDEP 2013). Cement kiln Hg emissions have subsequently increased through 2014 and represent 17% of the total state-wide Hg emissions. Coal-fired electric generation continues to be the largest Hg source category at 36% in Florida with most of these sources present in the northern and north-western parts of the state.

5.4 Mercury Sources and Emissions in the Rest of the United States and the World

The atmospheric modeling performed during the Florida state-wide Hg TMDL (FDEP 2013) showed that the overwhelming sources of Hg deposition in Florida are non-US in origin, transported by global weather patterns, and that only a very small percentage of Hg deposition occurring within Florida originated from within the state. As a result, we present here a review of Hg emissions from sources outside Florida including those in the rest of the world. In addition to emissions from anthropogenic sources, Hg emissions to the global atmosphere come from primary natural geogenic sources and the re-emission of historically-deposited Hg that was originally emitted from both anthropogenic and natural sources. The possible impacts of re-emissions of historically deposited or legacy Hg on future atmospheric deposition fluxes of Hg are discussed in Chap. 3 of Volume III.

5.4.1 Natural Sources and Re-emission of Mercury

Global primary natural (i.e., geogenic) sources of Hg include Hg released during geothermal activity in volcanoes and geothermal events and through the continuous natural weathering of Hg-containing rocks. Estimates of annual releases to air from natural sources worldwide vary widely, from 80 to 600 Mg/yr (Corbitt et al. 2011; Mason et al. 2012; Lei et al. 2014). Hg released from natural sources is primarily in the form of GEM which is either dry deposited or oxidized to GOM in the atmosphere before wet and dry deposition to the Everglades.

Re-emissions of historically deposited Hg occur from the Earth's surfaces (soil, rocks, snow and ice, oceans and other surface waters, and vegetation) that have previously received Hg either from atmospheric deposition or through another transport pathway (AMAP/UNEP 2013; Pacyna et al. 2016). Such Hg reflects the legacy contributions of both anthropogenic and natural sources and is re-emitted from the Earth's surfaces in the form of GEM. Oceans are the most important sources of these emissions and are estimated to be 52% of the total natural and re-emitted emissions (Pirrone et al. 2010) and 36% of the total global inventory when including direct anthropogenic sources (Pacyna et al. 2016). Because Hg evasion from ocean waters is significantly higher near the tropics than in temperate latitudes (Strode et al. 2007), evasion from the Atlantic Ocean of previously-deposited Hg potentially constitutes an important source of Hg to the Everglades. Total annual Hg re-emissions to the atmosphere of historically deposited Hg worldwide have been estimated to be in the range of 4000–6300 Mg/yr (Mason et al. 2012). The second largest source of Hg re-emissions is agricultural burning and wildfires where Hg stored in biomass is released as GEM and PBM. Estimates in the literature for annual global Hg emissions from biomass burning in the 2000s compare favorably with values reported at 675 ± 240 Mg/yr (Friedli et al. 2009), 600–678 Mg/yr (De Simone et al. 2015), and 612 Mg/yr (Kumar et al. 2018). Hg emissions from biomass burning in the United States have been estimated to be 43 Mg/yr on average during 2002–2006 (Wiedinmyer and Friedli 2007).

Global estimates of total primary natural emissions and re-emissions of Hg continue to remain uncertain with the combined estimates varying from 5207 Mg/yr (Pirrone et al. 2010; Pacyna et al. 2016) to 6500 Mg/yr (Lei et al. 2014). In any case, direct natural emissions and re-emissions worldwide constitute over two-thirds of the global Hg emission budget (Pacyna et al. 2016) and thus together represent an important source of worldwide Hg emissions.

5.4.2 Anthropogenic Mercury Emissions in the Rest of the United States and the World

Numerous inventories of global anthropogenic Hg emissions have been presented in the literature over the years with improvements made over time by considering more sources or applying more robust methods (e.g., Pacyna et al. 2003, 2006, 2010, 2016; Seigneur et al. 2004; Streets et al. 2009; Pirrone et al. 2010; AMAP/UNEP 2013). The increased application of Hg emission controls over the past two to three decades has slowed down or, in some cases, decoupled emissions from increases in energy demand, resulting in decreases in Hg emissions in Europe and North America. Conversely, a general increase in emissions in Asia, particularly in China and India, has been observed, resulting from increased industrial production and energy demands (Sundseth et al. 2017). The global geospatial distribution of anthropogenic Hg emissions reflects areas of industrialization (including known locations of point sources of Hg) and human population densities or remote extraction operations of metals (Fig. 5.7, from Pacyna et al. 2016).

Global anthropogenic Hg emissions are dominated by emissions from east and southeast Asia which represent 40% of the estimated 2010 worldwide inventory of 1960 Mg/yr (Fig. 5.8, prepared from data in AMAP/UNEP 2013). These emissions are primarily due to coal combustion, gold mining, cement production and non-ferrous metal (aluminium, copper, lead and zinc) production. North America contributes only 3% of the global anthropogenic total while Central America and the Caribbean contribute 2%. When considering all geographic regions in the world, artisanal and small scale gold mining represent the largest global mercury source sector,

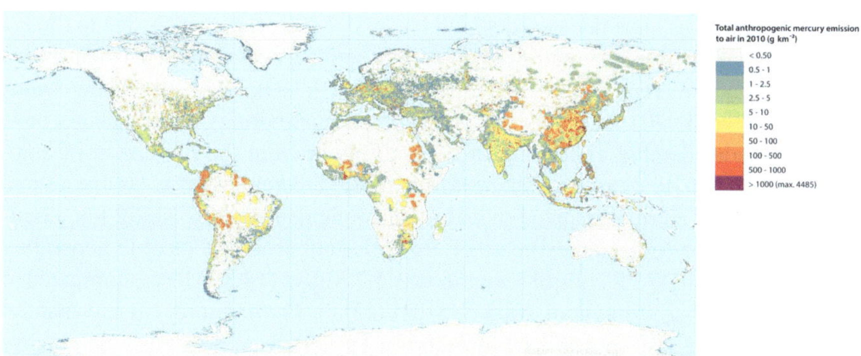

Fig. 5.7 Spatial distribution of global anthropogenic Hg air emissions in 2010. (From Pacyna et al. 2016)

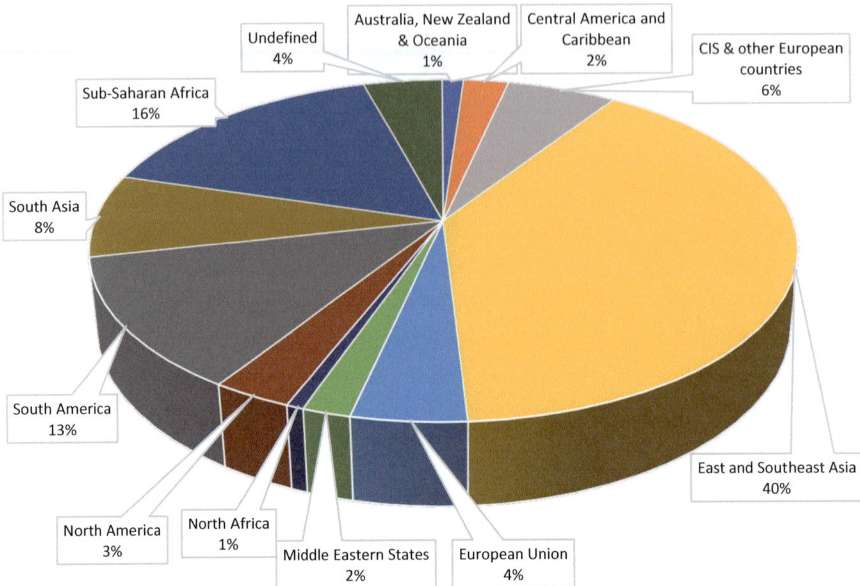

Fig. 5.8 Contribution of global anthropogenic Hg emissions in 2010 from different regions. (Figure created from data in AMAP/UNEP 2013)

constituting 37% of the total, followed by coal combustion at 24%. Other major emission sectors include production of nonferrous metals (10%), cement production (8.8%), and landfill waste and large-scale gold production at 5% each. There is significant uncertainty in the total estimate of global Hg emissions with an uncertainty range of 1010–4070 Mg/yr (AMAP/UNEP 2013; Pacyna et al. 2016) (Table 5.4).

The most recent U.S. national inventory of Hg emissions is from the 2014 USEPA National Emissions Inventory (NEI; USEPA 2014). Coal-fired power plants represent the largest anthropogenic source category of Hg emissions in the U.S., accounting for 44% of all domestic anthropogenic Hg emissions in 2014 (Table 5.5). U.S. Hg emissions have decreased significantly since the early 1990s due to the Hg co-benefits of the implementation of NOx, SO_2 and particulate matter (PM) source emissions controls for the implementation of USEPA NOx, SO_2 and PM emissions standards (FDEP, 2013). Emissions of Hg from coal-burning power plants have decreased considerably either due to Hg controls or from the retirement of coal-fired units in anticipation of implementation of the federal Mercury Air Toxics and Standards (MATS) rule finalized in 2012 (for implementation in 2016). For example, the USEPA's projected estimate of Hg emissions from US electric generating units in the 2016 MATS control scenario was 6.7 Mg/yr (14,800 lb/yr) compared to 47.8 Mg/yr (105,200 lb/yr) in 2005 (USEPA 2011). Furthermore, cheaper natural gas prices in the last decade have resulted in large transitions across the country from coal to natural gas usage for electric generation resulting in additional conversion or retirement of coal-fired power plants, thus further lowering Hg emissions due to the significantly lower content of Hg in pipeline natural gas versus coal.

Table 5.4 Global anthropogenic Hg air emissions by sector in 2010

Sector	Mg/yr	%
Artisanal and small-scale gold mining	727	37
Coal combustion	474	24
Non-ferrous metal production	194	10
Cement production	173	9
Large-scale gold production	97.3	5
Waste from consumer products (landfill)	89.4	5
Contaminated sites	82.5	4
Primary pig iron production	45.5	2
Chlor-alkali industry (Hg cell)	28.4	1.5
Oil refining	16.0	0.8
Mercury production	11.7	0.6
Oil and natural gas combustion	9.9	0.5
Waste from consumer products (incineration)	6.2	0.3
Cremation	3.6	0.2
Total	1959	100%

From AMAP/UNEP (2013)

Ferrous metals production, cement manufacturing and waste disposal comprise the next three largest source groups at 12%, 7%, and 7%, respectively, of the total U.S. anthropogenic inventory in 2014.

5.5 Source Contributions from Atmospheric Deposition to the Everglades

The Florida state-wide Hg TMDL (FDEP 2013) is the most comprehensive Hg modeling and monitoring study performed to date to understand atmospheric Hg deposition contributions to the Everglades and other parts of Florida. The methods and conclusions from this study are summarized below followed by other relevant studies.

5.5.1 Mercury Deposition Source Attribution from Florida State-Wide TMDL Study

The Florida state-wide Hg TMDL (FDEP 2013) included a rigorous atmospheric modeling effort supplemented with Hg deposition and air concentration monitoring by a team led by the University of Michigan Air Quality Laboratory (UMAQL) to characterize sources contributing to Hg deposition in Florida. An advanced multi-scale modeling system was applied consisting of the global ECHMERIT

Table 5.5 Anthropogenic Hg emissions by sector in the United States in 2014. From the 2014 NEI (USEPA 2014)

Sector	Emissions (lb/yr)	%
Fuel combustion – Electric generation – Coal	45,912.9	44
Industrial processes – Ferrous metals	12,103.6	12
Industrial processes – Cement manufacturing	7384.5	7
Waste disposal	6774.4	7
Miscellaneous Non-industrial NEC[a]	6235.3	6
Industrial processes – NEC[a]	4515.4	4
Industrial processes – Non-ferrous metals	2681.7	3
Industrial processes – Chemical Manufacturing	2085.3	2
Fuel combustion – Electric generation – Natural gas	1977.4	2
Fuel combustion – Industrial boilers, engines – Coal	1665.0	2
Fuel combustion – Residential – Oil	1287.5	1.2
Fuel combustion – Industrial boilers, engines – Natural gas	1244.7	1.2
Fuel combustion – Other	3910.3	3.8
Industrial processes – Petroleum refineries	1232.3	1.2
Mobile – Locomotives	1212.1	1.2
Industrial processes – Storage and transfer	769.9	0.7
Mobile – On-road	726.0	0.7
Industrial processes – Mining	531.1	0.5
Industrial processes – Oil and gas production	434.4	0.4
Industrial processes – Pulp and paper	368.6	0.4
Mobile – Non-road	114.0	0.1
Solvent – Industrial surface coating and solvent use	288	0.3
Agriculture – Livestock waste	95.7	0.1
Other	75.9	0.1
Total	103,626	

[a]*NEC* Not easily classifiable

(Atmospheric General Circulation Model ECHAM5 with Mercury Chemistry) model and regional CMAQ (Community Multiscale Air Quality) model. ECHMERIT (Jung et al. 2009) and CMAQ (Bullock and Brehme 2002; Byun and Schere 2006; Bullock et al. 2008, 2009; Baker and Bash 2012) are both Eulerian tropospheric chemistry transport models that simulate Hg emissions, transport, atmospheric chemistry and deposition. ECHMERIT was applied at 2.81°× 2.81° horizontal resolution for calendar year 2009 using the global AMAP/UNEP Hg inventory for year 2000, the USEPA NEI Hg inventory for year 2005, primary natural and biomass burning emissions, and meteorology for year 2009. The ECHMERIT simulation provided boundary conditions (background concentrations) of GEM, GOM and PBM for CMAQ simulations conducted at 36 km resolution over the continental U.S., 12 km resolution over the southeastern U.S., and 4 km resolution over Florida. The CMAQ modeling was performed for year 2009 using meteorology driven by the Weather Research and Forecast (WRF) for 2009 and emissions updated to year 2009 for Florida using data from the 2005/2008 USEPA NEI and 2009 emissions data from Florida DEP. The air modeling was supplemented with monitoring in 2009 and 2010 for precipitation, Hg wet deposition, Hg dry deposition and air concentrations of GEM, GOM and PBM. In addition, multivariate receptor modeling was conducted using the positive matrix factorization (PMF) technique (Paatero and Tapper 1994).

The CMAQ modeling (FDEP 2013) provided speciated wet and dry loading estimates of Hg for TMDL eco-regions and basins across the state of Florida for year 2009. It also provided a measure of source attribution, i.e., the relative importance of in-state versus out of state emissions for TMDL eco-regions and basins in Florida, as well as a measure of potential reduction in atmospheric loadings resulting from future emission reduction scenarios. Multiple tags were incorporated in the CMAQ modeling to identify relative source contributions of: (1) Florida sources; (2) states in the southeastern U.S. other than Florida; and (3) other source regions. The PMF receptor modeling highlighted potential source groups contributing to deposition.

The monitoring in the TMDL study showed that Hg rain concentrations in summertime (June – August) in the Everglades were 2–4 times higher than in other seasons (Fig. 5.6). Similarly, summertime Hg wet deposition is also typically higher than in the other seasons. Annual Hg wet deposition fluxes simulated by CMAQ for 2009 were compared with observed wet deposition at the Everglades National Park and five other TMDL monitoring stations in Florida (see Fig. 5.5 for station locations). The ENP had the highest observed annual wet deposition across all six TMDL stations in the state with a measured wet deposition flux of 25.5 μg/m^2-yr. The CMAQ model performance at ENP was very good with a modeled annual wet deposition flux of 25.3 μg/m^2-yr. Modeled dry deposition fluxes of Hg from CMAQ for year 2009 were evaluated by comparison with dry deposition estimates determined using a separate inferential model based on surface measurements of ambient speciated Hg and meteorology measured at four TMDL project "supersites" located at Davie, Tampa, Jacksonville, and Pensacola. Annual Hg dry deposition modeled with CMAQ (31 μg/m^2-yr) at Davie, the supersite closest to the ENP, compared very well with dry deposition flux estimate from the inferential modeling (32 μg/m^2-yr).

Comparisons between CMAQ model results and measurements of GEM, GOM and PBM were also conducted for calendar year 2009 for the four TMDL supersite locations (Davie, Tampa, Jacksonville and Pensacola). The model shows very reasonable agreement with measured GEM with overall normalized mean bias of +15% and normalized mean error of 16%. However, the model severely overpredicts GOM and PBM concentrations by a factor of 8–15 for both species. FDEP (2013) attribute this over-estimation to the model's inability to capture the decrease in GOM and PBM concentrations at night due to potential errors in emissions speciation, Hg chemical conversion, or removal processes. Other atmospheric models have tended to overestimate GOM and PBM concentrations as well (FDEP 2013). Nonetheless, the reasonably good performance of the TMDL CMAQ modeling for wet and dry deposition provides confidence in the estimates of Hg loading to the Everglades.

The Hg source apportionment scheme applied in the TMDL CMAQ modeling analysis showed that over 99% of the 25 μg/m^2-yr wet deposition flux at ENP was due to non-Florida sources, with most of that from the global background (FDEP 2013). Similarly, the modeling showed that only 1% of the approximately 17 μg/m^2-yr dry deposition flux at ENP was due to Florida sources. The total (i.e., wet +

dry) Hg deposition at ENP in 2009 was 42 μg/m^2-yr. Of this amount, less than 1% was contributed by Florida sources and over 99% was due to sources outside the state, primarily from outside the U.S.

The PMF analysis of Hg wet deposition data at the ENP site identified four significant factors contributing to Hg wet deposition (FDEP 2013). The most important factor was identified as a combined oil combustion/industrial source contributor, with others representing a crustal source, a marine source, and waste incineration.

In summary, the Florida state-wide Hg TMDL study concluded that Hg wet deposition is especially important in Florida because of the high frequency of convective storms (thunderstorms) in the Everglades and other parts of Florida, especially in summertime, and due to the large size of these weather systems. Convective storms can climb more than 16 km which allows the removal of GOM that has been formed from GEM by oxidation in the free troposphere after long-range transport from outside Florida; thus, the wet deposition often represents the Hg in a very large volume of the atmosphere (FDEP 2013). Additionally, thunderstorms can produce high speed winds pulling in still more volumes of air vertically from which the rain, or hail, removes atmospheric pollutants. Also, as noted below, thunderstorms may increase Hg concentrations in non-convective precipitation by mixing GOM down to lower altitudes where it can be scavenged by all precipitation types. Across Florida, thunderstorms are more common in inland areas by about 20%, and across coasts to inland areas thunderstorms occur on average of 80–100 days per year (FDEP 2013). The scale of rain from summertime thunderstorms in the Everglades often exceeds 7.6 cm in an hour and enables Hg to be scavenged rapidly from the atmosphere.

5.5.2 Mercury Deposition Source Attribution from Other Modeling and Monitoring Studies

Over the past three decades numerous other modeling and monitoring studies have attempted to explain the high wet deposition observed in the Everglades and the rest of southern Florida and to identify the relative contributions of various emission source groups and regions to both wet and dry deposition of Hg in the region.

As noted previously (Chap. 4, this volume), a hybrid air quality modeling approach was used in the Everglades Hg Pilot TMDL Study (FDEP 2003) to show that dry deposition was an important component of Hg deposition in the Everglades during June 1995 to June 1996, comprising 34–40% of the total Hg deposition. Dry deposition was dominated by the GOM fraction and showed a seasonal trend, with relatively greater deposition occurring during the climatological wet season. FDEP (2003) also concluded that the relatively modest decreases in measured volume-weighted mean Hg concentrations in rainfall from 1993 to 2000 at the Everglades National Park Beard Research Center agree reasonably well with the Hg emissions declines in Dade, Broward and Palm Beach counties during that time period. Two key issues relevant to Hg deposition were not addressed in the Pilot TMDL study and were recommended for follow-on work: (1) the inclusion of the global-scale contribution of Hg to deposition in the Everglades was considered outside the scope of the study due to paucity of data or models at the time; and (2) Hg species transformations in the atmosphere needed to be better characterized (FDEP 2003).

The FAMS study (see Chap. 4, this volume) allowed for the estimation of Hg deposition throughout Florida during 1992–1996 using precipitation and concentration monitoring (Guentzel et al. 2001). Most of the Hg wet deposition occurred during the summertime and Hg concentrations in precipitation were found to be significantly higher in southern Florida than at the north-central and north-western sites. Also, Hg concentrations in precipitation in southern Florida were similar at the urban and rural sites. In contrast, concentrations of related trace metals such as V, Ni, Cu, Zn, and Pb in precipitation were 20–80% higher at the urban sites. Guentzel et al. (2001) concluded that Hg was scavenged from the global pool in the free troposphere. They used box model calculations to hypothesize that long-range transport of GOM (referred to as reactive gaseous mercury or RGM by the authors), coupled with strong convective thunderstorm activity during the summertime, represents >50% of the Hg deposition in southern Florida. Their box model calculations also suggested that local anthropogenic GOM and particulate Hg emissions accounted for 30–46% of the summertime rainfall Hg deposition across southern Florida peninsula during 1992–1996 and the remainder from the global background.

The South Florida Atmospheric Mercury Monitoring Study (SoFAMMS) was a short-term study conducted from 6 August to 6 September 1995 to examine the potential impacts of local anthropogenic sources on Hg in precipitation in southern Florida using highly resolved sampling in both space and time (Dvonch et al. 1998). The study hypothesized that the large spatial and temporal variations in observed Hg concentrations in precipitation, which could not be explained by precipitation depth alone, indicated large impacts from local sources. As part of subsequent analyses, daily event precipitation sample data were used with a receptor modeling approach to investigate sources of Hg wet deposition in south Florida. A multivariate receptor modeling approach based on principal component factor analysis (PCA) estimated that municipal waste incineration accounted for 57% and oil combustion for 14% of the Hg wet deposited at five Florida Everglades sites located near or just within the eastern edge of the Everglades Protection Area (EvPA) (Dvonch et al. 1999). Dvonch et al. (2005) studied the influence of meteorological conditions on the wet deposition of Hg in south Florida using data collected in Davie, Florida during June 1995 to June 1996. A meteorological tracer analysis, utilizing the ratio of trace elements lanthanum to cerium as a tracer of oil-fired combustion emissions, determined that feed air to precipitation cells arriving at the site incorporated local urban emissions more frequently during the spring and summer seasons. The authors concluded that local anthropogenic sources played a dominant role in the wet deposition of Hg to southern Florida during the period investigated.

Lin et al. (2005) adapted the Biogenic Emission Inventory System (BEIS) for estimating Hg emissions from vegetation across the U.S. using land cover data, modeled meteorology and emission factors. The modeled Hg emissions were evaluated by comparison with evasion fluxes measured in Florida wetlands in June 1997

(Lindberg et al. 2002). The total vegetative Hg emissions in the continental US domain were estimated to be 44 ton/yr during 2001. They were highest in summer and were mainly contributed from the southeastern US. In Florida, they are highest in the southern (including the Everglades) and northwestern parts of the state.

Seigneur et al. (2004) used a nested multiscale modeling system consisting of a global chemical transport model (CTM-Hg) at 8° × 10° horizontal resolution and a continental chemical transport model (the Trace Element Analysis Model, TEAM) at 80–100 km horizontal resolution to simulate the chemistry, transport and deposition of anthropogenic, natural and re-emitted Hg over the continental US in 1998 (the horizontal resolution of TEAM varies with latitude and was, in this application, ~84 km over southern Florida). The models were evaluated against MDN wet deposition data and ambient Hg concentrations and subsequently used to conduct a source attribution analysis for 19 ecologically-sensitive receptor areas including three in Florida. The modeled contribution of North American anthropogenic emissions to Hg deposition at the Everglades National Park is 20%. The estimated contribution of natural sources at the ENP is 25% and the contribution of Hg emissions from anthropogenic sources in other continents exceeds 50%. Carlton et al. (2004) reported results from an application of the same model, TEAM, at a finer horizontal resolution (15–20 km). U.S. anthropogenic sources were modeled to contribute 8% to deposition in the Everglades in 2004. The USEPA (USEPA 2005) applied the CMAQ model at 36 km horizontal resolution with boundary conditions from GEOS-Chem for 2001 as part of Clean Air Mercury Rule (CAMR) modeling. They determined that global sources (i.e., those outside North America) contribute >85% to Hg deposition in southern Florida.

During the Florida Everglades Dry Deposition Study (FEDDS), Marsik et al. (2007) made measurements of Hg dry deposition during February – March 1999 and June 2000 using a surrogate water surface technique and compared these with modeled estimates of dry deposition using a single-layer inferential model. The average daily dry deposition fluxes measured during the 1999 and 2000 periods were 13.3 and 5.9 ng/m^2-day, respectively, while the modeled fluxes were lower, 3.4 and 1.8 ng/m^2-day, respectively. The authors hypothesize that the reductions in dry deposition from 1999 to 2000 could be due to reductions in waste incineration emissions in the interim period and/or to more wet removal in the summer period of 2000 compared to the winter of 1999. The authors note that the dry deposition of Hg is likely to be important across south Florida during the months of October–May, when precipitation is relatively infrequent. They attribute the discrepancy between measured and modeled estimates of Hg dry deposition to one or more of the following reasons: (1) errors in the measurement of ambient concentrations of GOM and PBM; (2) uncertainties in the surrogate water surface method employed to measure dry deposition; and (3) inaccuracies in the description of the resistances to uptake of GOM and PBM by the canopy.

Selin et al. (2007) applied GEOS-Chem at 4° × 5° horizontal resolution to simulate Hg deposition in North America and other continents and successfully validated the model against observed Hg air concentrations and MDN wet deposition data in 2003. They attribute the observed maximum MDN wet deposition flux in the southeastern U.S. to photochemical oxidation (by ozone or hydroxyl (OH) radicals) of the global GEM pool and frequent precipitation. They estimate that North American anthropogenic sources contribute, on average, less than 10% to wet plus dry deposition in southern and central Florida. Selin and Jacob (2008) applied an updated version of GEOS-Chem to simulate Hg deposition in 2004–2005. The model reproduced well measured Hg air concentrations and both the seasonal cycle and latitudinal gradient in the MDN wet deposition data (maximum in summer and increasing in amplitude from north to south). The authors interpret the seasonal cycle in the MDN data as largely driven by the global pool of Hg and attribute the high observed summertime Hg wet deposition in the Southeast (and, in particular, Florida) to the interaction of global-scale subtropical downwelling, which supplies elevated divalent mercury (GOM) in subsiding air masses, with frequent regional deep convection, which scavenges this free tropospheric GOM from high altitudes. They estimate that North American anthropogenic sources contribute, on average, 10–15% to wet + dry deposition in southern Florida. Vijayaraghavan et al. (2007) report that observed Hg wet deposition fluxes show a clear north-to-south gradient across the U.S. with greater deposition in Florida. This pattern is unlike observed sulfate wet deposition fluxes that show larger values in the northeast (near large emission sources) than the southeast, thus suggesting that sulfate wet deposition is more strongly influenced by local/regional emission sources than Hg.

There is an increased need for improved ambient measurements, including Hg speciation, as well as model sophistication towards reconciling model simulations and observed Hg chemistry and deposition. Sillman et al. (2007) applied CMAQ with modifications to include an integrated solution for gas phase and aqueous Hg photochemistry and expanded atmospheric chemistry over a model domain with 36 km horizontal resolution over the eastern half of the U.S. and parts of Canada, Mexico and the Caribbean. The model was applied for 15 days in June 2000 and reproduced the overall patterns in aircraft measurements in south Florida that show GOM varying between 10 and 230 pg/m^3 in the troposphere and increasing with altitude. However, the authors note that the model under-predicts the maximum observed GOM by a factor of two and, that if the reservoir of elevated GOM in the upper troposphere predicted by Selin et al. (2007) had been included here, this might have resulted in higher modeled GOM that would be closer to the observations. Gridded Hg models such as CMAQ dilute the mass of Hg emitted over the volume of each model grid cell. Vijayaraghavan et al. (2008) reported better model predictions with CMAQ for Hg wet deposition at MDN monitoring stations in the U.S. when applying a plume-in-grid treatment (advanced plume treatment or APT) that initially models the chemistry and transport of Hg in the puffs released from elevated point sources before transferring the mass to the grid farther away from the source. Baker and Bash (2012) modeled Hg deposition over the U.S. at 12 km horizontal resolution with CMAQ and the Comprehensive Air Quality Model with extensions (CAMx) (Ramboll 2018) and concluded that speciated ambient observations that provide more insight into air-surface exchange and an increased understanding of atmospheric Hg chemical processes will further constrain model parameterizations and improve model performance.

Using a box model for the marine boundary layer (MBL), Holmes et al. (2009) concluded that a major source of Hg deposition in the MBL is from the oxidation of GEM transported in the free troposphere over long distances to GOM by bromine atoms and subsequent deposition after uptake onto sea-salt aerosols. The incorporation of a halogen chemical mechanism that includes bromine-induced gas-phase oxidation of GEM in CMAQ by Ye et al. (2018) resulted in improved model predictions of observed GOM and PBM concentrations and wet and dry deposition in the eastern U.S. Through measurements of vertical profiles of bromine monoxide (BrO) radicals, GEM and GOM over the Gulf coast of Florida, Coburn et al. (2016) hypothesize that the oxidation by bromine radicals is the dominant pathway for GEM oxidation to form water-soluble GOM (with less than a 5% increase due to chlorine-induced oxidation) that is subsequently available for wet scavenging by thunderstorms or transport to the boundary layer. Thus, the literature indicates that the oxidation of GEM aloft is driven by a larger natural than anthropogenic component.

Holmes et al. (2016) used rainwater samples from over 800 individual precipitation events and radar and satellite observations at seven locations across the eastern U.S., including one in the Florida panhandle, to show that thunderstorms increase Hg concentrations by 50% relative to weak convective or stratiform events of equal precipitation depth due to the strong convection reaching the upper troposphere where GOM resides. The authors also note that thunderstorms may also raise baseline Hg concentrations in non-convective precipitation by mixing GOM down to lower altitudes where it can be scavenged by all precipitation types. Using aircraft measurements of GOM and total Hg air concentrations in 2013 and the GEOS-Chem model, Shah et al. (2016) showed that high summertime GOM concentrations in the free troposphere in the southeastern US were associated with clean subsiding air masses originating in the upper troposphere within the Pacific or Atlantic anticyclones. The anticyclones are characterized by large-scale sinking motion which transports higher GOM concentrations (> 300 pg/m^3) from the upper troposphere due to the fast oxidation of GEM resulting from higher Br concentrations. The sinking air in the anticyclones also suppresses cloud formation and precipitation, thereby preventing loss of GOM by reduction and wet deposition (Shah et al. 2016). The authors conclude therefore that the transport of GOM produced in the Pacific anticyclone could be an important source of GOM deposition over the southeastern U.S. if these high GOM air masses are exposed to deep convection.

5.6 Conclusions

In summary, the historical Hg deposition to the Everglades until the mid-1990s was likely due to a combination of emissions from local sources with a high oxidized Hg speciation fraction (e.g., >75%) and short stacks such as waste incinerators, as well as regional and more distant sources. Local anthropogenic GOM and particulate Hg emissions accounted for 30–46% of the summertime rainfall Hg deposition across southern Florida peninsula during 1992–1996 with the remainder from the global background. Despite reductions in local emission sources since that time, the continuing high deposition of Hg to the Everglades appears to be largely due to the oxidation of gaseous elemental Hg in the free troposphere (originating from regional and global emission sources) to water-soluble gaseous divalent Hg by bromine and other oxidants followed by: (1) scavenging during deep convective activity and subsequent wet deposition especially during summertime thunderstorms in the Everglades; or (2) transport to the boundary layer and subsequent dry deposition or wet deposition during non-convective precipitation. By the late 1990s and early 2000s, due to local source emission reductions, over 99% of the 25 μg/m^2-yr wet deposition flux at ENP was due to non-Florida sources, with most of that from the global background (FDEP 2013). Similarly, computer modeling from the Florida Hg TMDL study (FDEP 2013) showed that only 1% of the approximately 17 μg/m^2-yr dry deposition flux at ENP was due to Florida sources. The total (i.e., wet + dry) Hg deposition at ENP in 2009 is 42 μg/m^2-yr. Of this amount, less than 1% is contributed by Florida sources and over 99% is due to sources outside the state, primarily from outside the U.S.

More research is needed to identify missing Hg emission sources, if any, in current southern Florida emission inventories and to improve our understanding of Hg dry deposition velocities and atmospheric Hg chemistry and scavenging over the Everglades, all of which affect the analysis of Hg deposition and biogeochemical cycling in the region.

References

AMAP/UNEP (2013) Technical background report for the global mercury assessment 2013. Arctic Monitoring and Assessment Programme, Oslo, Norway/UNEP Chemicals Branch, Geneva, Switzerland. (www.amap.no/documents/download/1265)

Baker KR, Bash JO (2012) Regional scale photochemical model evaluation of total mercury wet deposition and speciated ambient mercury. Atmos Environ 49:151–162

Bullock OR, Brehme KA (2002) Atmospheric mercury simulation using the CMAQ model: formulation description and analysis of wet deposition results. Atmos Environ 36(13):2135–2146

Bullock OR Jr, Atkinson D, Braverman T, Civerolo K, Dastoor A, Davignon D, Ku JY, Lohman K, Myers TC, Park RJ, Seigneur C, Selin NE, Sistla G, Vijayaraghavan K (2009) An analysis of simulated wet deposition of mercury from the North American Mercury Model Intercomparison Study. J Geophys Res 114:D08301

Bullock OR, Atkinson D, Braverman T, Civerolo K, Dastoor A, Davignon D, Ku JY, Lohman K, Myers TC, Park RJ, Seigneur C, Selin NE, Sistla G, Vijayaraghavan K (2008) The North American mercury model intercomparison study (NAMMIS): study description and model-to model comparisons. J Geophys Res Atmos 113:1–17

Byun D, Schere KL (2006) Review of the governing equations, computational algorithms, and other components of the models-3 Community Multiscale Air Quality (CMAQ) modeling system. Appl Mech Rev 59:51–77

Carlton R, Chu P, Levin L, Offen G, Yager J (2004) EPRI comments on EPA proposed performance, electric utility steam generating units: mercury emissions – national emission standards for hazardous air pollutants; and, in the alternative, proposed standards of performance for new and existing stationary sources: electric utility steam generating units; proposed rule (Federal register 69, 20, January 30, 2004). Docket ID No. OAR-2002-0056. EPRI, Palo Alto, CA

Coburn S, Dix B, Edgerton E, Holmes CD, Kinnison D, Liang Q et al (2016) Mercury oxidation from bromine chemistry in the free troposphere over the southeastern US. Atmos Chem Phys 16:3743–3760. https://doi.org/10.5194/acp-16-3743-2016

Corbitt ES, Jacob DJ, Holmes CD, Streets DG, Sunderland EM (2011) Global source-receptor relationships for mercury deposition under present-day and 2050 emissions scenarios. Environ Sci Technol 45:10477–10484

De Simone F, Cinnirella S, Gencarelli CN, Yang X, Hedgecock IM, Pirrone N (2015) Model study of global mercury deposition from. Biomass Burning Environ Sci Technol 49(11):6712–6721

Dvonch JT, Graney JR, Marsik FJ, Keeler GJ, Stevens RK (1998) An investigation of source–receptor relationships for mercury in south Florida using event precipitation data. Sci Total Environ 213:95–108

Dvonch JT, Graney JR, Keeler GJ, Stevens RK (1999) Use of elemental tracers to source apportion mercury in South Florida precipitation. Environ Sci Technol 33(24):4522–4527

Dvonch JT, Keeler GJ, Marsik FJ (2005) The influence of meteorological conditions on the wet deposition of mercury in Southern Florida. J Appl Meteorol 44:1421–1435

FDEP (Florida Department of Environmental Protection) (2003) Integrating atmospheric mercury deposition and aquatic cycling in the Florida Everglades: an approach for conducting a Total maximum daily load analysis for an atmospherically derived pollutant. Final report. Authors: Atkeson T, Axelrad D, Pollman CD, Keeler G. Prepared for United States Environmental Protection Agency. October

Florida Department of Environmental Protection (FDEP) (2013) Mercury TMDL for the State of Florida. Final report. Watershed evaluation and TMDL Section. October 24. https://floridadep.gov/sites/default/files/Mercury-TMDL.pdf

Friedli H, Arellano A, Cinnirella S, Pirrone N (2009) Initial estimates of mercury emissions to the atmosphere from global biomass burning. Environ Sci Technol 43:3507–3513

Guentzel JL, Landing WM, Gill GA, Pollman CD (2001) Processes influencing rainfall deposition of mercury in Florida. Environ Sci Technol 35:863–873

Holmes CD, Jacob DJ, Mason RP, Jaffe DA (2009) Sources and deposition of reactive gaseous mercury in the marine atmosphere. Atmos Environ 43:2278–2285

Holmes CD, Krishnamurthy NP, Caffrey JM, Landing WM, Edgerton ES, Knapp KR, Nair US (2016) Thunderstorms increase mercury wet deposition. Environ Sci Technol 50:9343–9350. https://doi.org/10.1021/acs.est.6b02586

Husar JD, Husar RB (2002) Trends of anthropogenic mercury mass flows and emissions in Florida. Final report, prepared for Florida Department of Environmental Protection under subcontract to TetraTech Inc. Lantern Corporation, Clayton, MO

Jung G, Hedgecock IM, Pirrone N (2009) ECHMERIT V1.0 – a new global fully coupled mercury-chemistry and transport model. Geosci Model Dev 2:175–195

Kumar A, Wu S, Huang Y, Liao H, Kaplan JO (2018) Mercury from wildfires: global emission inventories and sensitivity to 2000–2050 global change. Atmos Environ 173:6–15

Lei H, Wuebbles DJ, Liang X-Z, Tao Z, Olsen S, Artz R, Ren X, Cohen M (2014) Projections of atmospheric mercury levels and their effect on air quality in the United States. Atmos Chem Phys 14(2):783–795

Lin C-J, Lindberg SE, Ho TC, Jang C (2005) Development of a processor in BEIS3 for estimating vegetative mercury emission in the continental United States. Atmos Environ 39:7529–7540

Lindberg SE, Dong W, Myers T (2002) Transpiration of gaseous elemental mercury through vegetation in a subtropical wetland in Florida. Atmos Environ 36:5207–5219

Marsik FJ, Keeler GJ, Landis MS (2007) The dry-deposition of speciated mercury to the Florida Everglades: measurements and modeling. Atmos Environ 41:136–149

Mason RP, Choi AL, Fitzgerald WF, Hammerschmidt CR, Lamborg CH, Soerensen AL, Sunderland EM (2012) Mercury biogeochemical cycling in the ocean and policy implications. Environ Res 119:101–117

Paatero P, Tapper U (1994) Positive matrix factorization: a non-negative factor model with optimal utilization of error estimates of data values. Environmetrics 5(2):111–126. https://doi.org/10.1002/env.3170050203

Pacyna JM, Pacyna EG, Steenhuisen F, Wilson S (2003) Mapping 1995 global anthropogenic emissions of mercury. Atmos Environ 37:109–117

Pacyna EG, Pacyna JM, Steenhuisen F, Wilson S (2006) Global anthropogenic mercury emission inventory for 2000. Atmos Environ 40:4048–4063

Pacyna EG, Pacyna JM, Sundseth K, Munthe J, Kindbom K, Wilson S, Steenhuisen F, Maxson P (2010) Global emission of mercury to the atmosphere from anthropogenic sources in 2005 and projections to 2020. Atmos Environ 44:2487–2499

Pacyna JM, Travnikov O, De Simone F, Hedgecock IM, Sundseth K, Pacyna EG, Steenhuisen F, Pirrone N, Munthe J, Kindbom K (2016) Current and future levels of mercury atmospheric pollution on global scale. Atmos Chem Phys 16:12495–12511

Pirrone N, Cinnirella S, Feng X, Finkelman RB, Friedli HR, Leaner J, Mason R, Mukherjee AB, Stracher GB, Streets DG, Telmer K (2010) Global mercury emissions to the atmosphere from anthropogenic and natural sources. Atmos Chem Phys 10:5951–5964

Ramboll (2018) User's guide – comprehensive air quality model with extensions (CAMx). Available at www.camx.com

RMB Consulting & Research, Inc. (RMB) (1999) South Florida mercury emissions inventory (1996). Final report. Prepared for Florida Department of Environmental Protection. Prepared by RMB Consulting & Research, Inc., Raleigh, NC

RMB Consulting & Research, Inc. (RMB) (2002) Atmospheric mercury emissions from major point sources – broward, dade, and Palm beach counties, 1980–2000. Final report. RMB Consulting & Research, Inc., Raleigh, NC

Seigneur C, Vijayaraghavan K, Lohman K, Karamchandani P, Scott C (2004) Global source attribution for mercury deposition in the United States. Environ Sci Technol 38:555–569

Selin NE, Jacob DJ (2008) Seasonal and spatial patterns of mercury wet deposition in the United States: constraints on the contribution from North American anthropogenic sources. Atmos Environ 42:5193–5204

Selin NE, Jacob DJ, Park RJ, Yantosca RM, Strode S, Jaeglé L, Jaffe D (2007) Chemical cycling and deposition of atmospheric mercury: global constraints from observations. J Geophys Res 112:D02308

Shah V, Jaeglé L, Gratz LE, Ambrose JL, Jaffe DA, Selin NE, Song S, Campos TL, Flocke FM, Reeves M et al (2016) Origin of oxidized mercury in the summertime free troposphere over the southeastern US. Atmos Chem Phys 16:1511–1530. https://doi.org/10.5194/acp-16-1511-2016

Sillman S, Marsik FJ, Al-Wali KI, Keeler GJ, Landis MS (2007) Reactive mercury in the troposphere: model formation and results for Florida, the northeastern United States, and the Atlantic Ocean. J Geophys Res 112:D23305. https://doi.org/10.1029/2006JD008227

Streets DG, Zhang Q, Wu Y (2009) Projections of global mercury emissions in 2050. Environ Sci Technol 43:2983–2988

Strode S, Jaegle L, Selin N, Jacob D, Park R, Yantosca R, Mason R, Slemr F (2007) Air-sea exchange in the global mercury cycle. Glob Biogeochem Cycles 21:GB1017

Sundseth K, Pacyna J, Pacyna E, Pirrone N, Thorne R (2017) Global sources and pathways of mercury in the context of human health. Int J Environ Res Public Health 14(1):105

USEPA (2005) Technical support document. Revision of December 2000 regulatory finding on the emissions of hazardous air pollutants from electric utility steam generating units and the removal of coal and oil-fired electric utility steam generating units from the section 112(c) list: reconsideration, U.S. Environmental Protection Agency, Washington, DC, October 21, 2005

USEPA (2011) Air quality modeling technical support document: EGU mercury analysis. EPA-454/R-11-008. Office of air quality planning and standards, U.S. environmental protection agency, September

USEPA (2014) 2014 National Emissions Inventory (NEI) Data. Available at https://www.epa.gov/air-emissions-inventories/2014-national-emissions-inventory-nei-data

Vijayaraghavan K, Seigneur C, Karamchandani P, Chen S-Y (2007) Development and application of a multi-pollutant model for atmospheric mercury deposition. J Appl Meteorol Climatol 46:1341–1353

Vijayaraghavan K, Karamchandani P, Seigneur C, Balmori R, Chen S-Y (2008) Plume-in-grid modeling of atmospheric mercury. J Geophys Res 113:D24305. https://doi.org/10.1029/2008JD010580

Wiedinmyer C, Friedli H (2007) Mercury emission estimates from fires: an initial inventory for the United States. Environ Sci Technol 41:8092–8098

Ye Z, Mao H, Driscoll CT, Wang Y, Zhang Y, Jaeglé L (2018) Evaluation of CMAQ coupled with a state-of-the-art mercury chemical mechanism (CMAQ-newHg-Br). J Adv Model Earth Syst 10(3):668–690

Chapter 6
Atmospheric Deposition of Mercury in the Everglades – Synthesis of Global Cycling and Local Process Effects

Curtis D. Pollman, Krish Vijayaraghavan, William M. Landing, and Christopher D. Holmes

Abstract This chapter synthesizes information on the global cycling of mercury (Hg) in the atmosphere (Chap. 3, this volume) with information on atmospheric deposition dynamics of Hg in the Everglades (Chap. 4, this volume) and the relationship of different atmospheric sources (as a function of spatial scale processes) with the Everglades as a receptor (Chap. 5, this volume). The interplay of all three of these components, and how local source impacts have changed over time, have profound implications regarding both understanding the Everglades Hg problem and the possibility of effectively mitigating the problem through controls on Hg emissions.

Keywords Source-receptor relationships · Emissions · Convective storm · Free troposphere · Gaseous oxidized mercury · Gaseous elemental mercury

6.1 Introduction

Atmospheric deposition fluxes of mercury (Hg) to the Florida Everglades are amongst the highest in the United States for background locations not located within urban areas. Atmospheric deposition is also the primary source of external inputs of Hg to the Everglades (>95%; USEPA 1996). On the surface, these two facts would suggest that mitigating the Hg problem in the Everglades could be resolved by understanding

C. D. Pollman (✉)
Aqua Lux Lucis, Inc, Gainesville, FL, USA
e-mail: cpollman@aqualuxlucis.org

K. Vijayaraghavan
Ramboll US Corporation, Novato, CA, USA

W. M. Landing · C. D. Holmes
Department of Earth, Ocean and Atmospheric Science, Florida State University, Tallahassee, FL, USA

C. D. Pollman et al. (eds.), *Mercury and the Everglades. A Synthesis and Model for Complex Ecosystem Restoration*,
https://doi.org/10.1007/978-3-030-20070-1_6

and controlling these fluxes. Indeed, early research into the Everglades Hg problem included studies to quantify the magnitude and sources of the Hg signal in the region, including fluxes in wet deposition (Pollman et al. 1995; Guentzel et al. 1995, 1998 and 2001); the amount of gaseous elemental Hg (GEM) in ambient air (Gill et al. 1995); evaluating local source-receptor relationships (i.e., the quantitative link between emission sources and affected locations) influencing Hg concentrations in wet deposition (Dvonch et al. 1999); and constructing inventories of local emissions (Husar and Husar 2002; RMB Consulting and Research 1999 and 2002). In part, this work eventually led to a pilot total maximum daily load (TMDL) study[1] funded by the USEPA and conducted through the Florida Department of Environmental Protection (FDEP). This TMDL study sought to define critical loadings of Hg protective of human health (where the exposure pathway was through the consumption of higher trophic level fish and largemouth bass in particular) and source allocation limits necessary to achieve those critical loadings (Atkeson et al. 2003).

6.2 Atmospheric Hg Cycling and Source Contributions

At a minimum Hg deposition fluxes at any given location reflect large spatial scale components driven by global cycling. These large-scale components include both natural sources (*e.g.,* geogenic emissions such as volcanic releases of Hg into the atmosphere) and anthropogenic sources (both direct emissions such as fossil fuel consumption and emissions of historically deposited [legacy] natural and anthropogenic Hg re-emitted from terrestrial and oceanic Hg pools). Natural emissions combined with re-emissions of legacy anthropogenic Hg contribute perhaps two-thirds or more to the global Hg budget, with direct anthropogenic emissions comprising the remainder.

Superimposed on these large-scale fluxes is a veneer of fluxes with localized (0–100 km) to regional components (100–1000 km). The magnitude of this veneer is related to the speciation of the Hg emitted from the source, the magnitude of the emissions, the proximity of the source to the receptor, and local meteorology. Hg can be emitted and is present in the atmosphere in any of three forms – GEM, particle bound oxidized Hg (PBM), and gaseous oxidized Hg (GOM) – each with different depositional dynamics. GEM reacts slowly in the atmosphere and is only sparingly soluble in water. As a result, it resides in the atmosphere on average for 6–12 months. PBM and GOM are easily scavenged from the atmosphere by rainfall, and GOM readily dry deposits to landscape surfaces because of its high reactivity. The atmospheric residence time of these two forms thus is on the order of only days

[1] This study was considered a pilot because it was the first TMDL study that explicitly considered inputs of a pollutant to an aquatic ecosystem originating from airborne emissions and delivered through atmospheric deposition, rather than through terrestrial (surface and ground water) pathways alone. As such the study used modeling of both atmospheric source-receptor relationships and the biogeochemical cycling of Hg in the receptor aquatic ecosystem to quantitatively link source emissions to biota response.

to weeks. When considering the impact of local emission sources on receptor aquatic ecosystems the question of Hg speciation in the emissions thus becomes critically important. GEM emissions will largely enter into the global pool and be redistributed over large spatial scales before finally being removed. PBM and GOM emissions will be rapidly removed and deposit largely within less than 100 km of the emission source. This is illustrated for a hypothetical GOM source of 1 kg/day as part of a series of simulations conducted by Cohen et al. (2007) to demonstrate the importance of emissions speciation on near-field deposition fluxes (Fig. 6.1). Restated, local and regional sources of GOM and PBM will have greater local and regional impacts on Hg deposition than sources of GEM.

The sources of Hg driving deposition in the Everglades currently are dominated almost entirely by regional and larger scale emissions. This is a direct consequence of low emission fluxes of GOM from local sources. Based on 2005 NEI estimates (USEPA 2005), GOM emissions total perhaps 94 kg/yr in the south coastal Florida urban fringe extending south from St. Lucie to Dade county (Fig. 6.2). The amount of atmospheric Hg directly deposited to the remnant Everglades (623,200 ha; see Chap. 1, this volume) is at least 179 kg/yr, based on annual average wet deposition

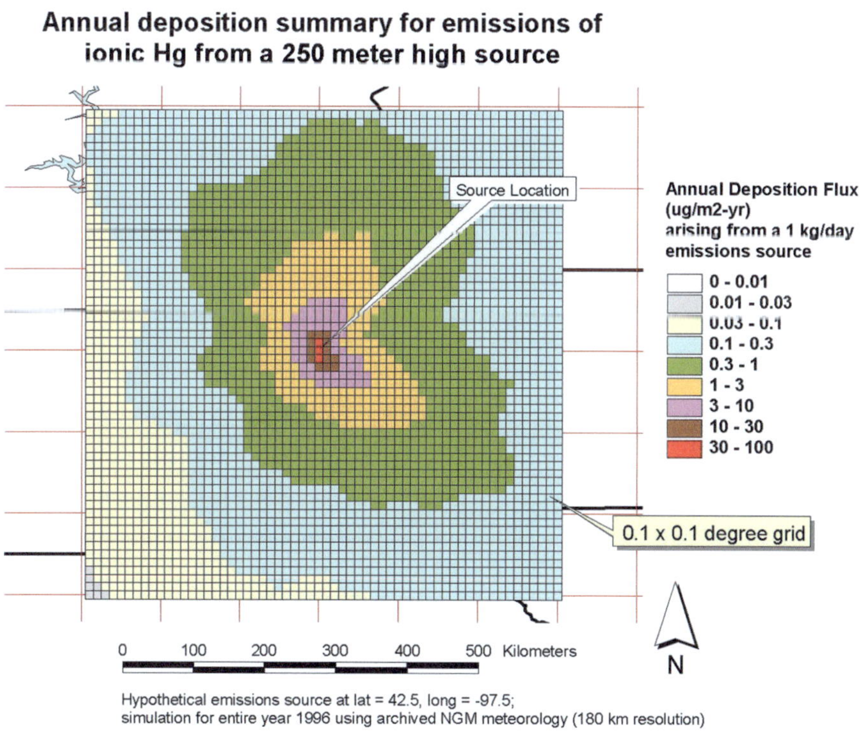

Fig. 6.1 Model analysis by Cohen et al. (2007) of the deposition field of GOM from a hypothetical emissions source with a 250 m tall stack and GOM emissions of 1 kg/day. (Reprinted with permission from the author)

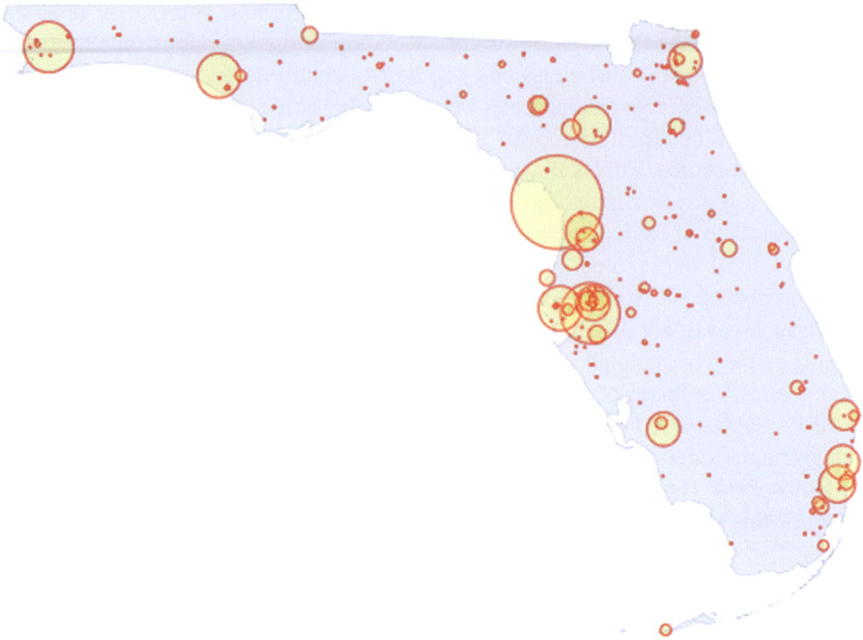

Fig. 6.2 GOM emissions in Florida, 2005. (Data from the National Emissions Inventory (USEPA 2005) for Hg. Circle areas are proportional to emissions; maximum emission is 209 kg/yr. These data were part of the emissions database used to conduct the source-receptor modeling as part of the Florida statewide Hg TMDL (FDEP 2013)

fluxes measured between 2007 and 2016 at the Mercury Deposition Network site located at the Beard Research Center in Everglades National Park, and assuming dry deposition equates to 50% the wet deposition flux. The magnitude of local GOM emissions thus corresponds numerically to 53% of the total Hg deposition flux to the Everglades; for local emissions to be a significant contributor to this flux thus requires that local emissions largely deposit in the Everglades. Because of distance and prevailing meteorology, however, only a fraction of the GOM emissions from the urban fringe reach and are deposited across the Everglades. This fact underlies the strong agreement across several different modeling studies indicating that local to statewide sources now contribute to less than 7% (and more likely ~ 3%) of the Hg signal in wet and dry deposition fluxing into the Everglades.

Historically, however, the contribution from localized sources likely was much more substantial and significant. This possibility is suggested by two separate and essentially contemporaneous inventories – each using a fundamentally different approach towards constructing the inventory – of local Hg emissions originating from south Florida urban fringe. The first inventory was constructed by Husar and Husar (2002) on an annual basis between 1930 and 2000 by estimating the amount of Hg released into the environment based on material usage. The second inventory

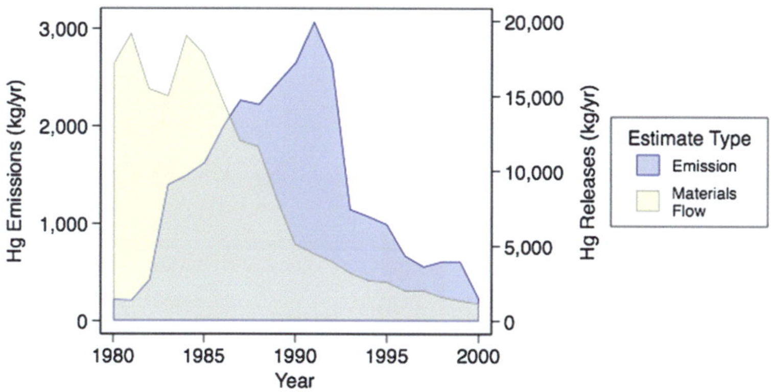

Fig. 6.3 Annual Hg emissions (RMB 2002) and estimated releases of Hg into the environment based on materials flow analysis for south Florida (Husar and Husar 2002) between 1980 and 2000

was based on emissions records and assumed Hg emission factors for different Hg sources (RMB 2002). Both inventories included analyses of annual fluxes specific to the south Florida urban fringe (Broward and Miami-Dade counties for both inventories, while the RMB inventory also included Palm Beach county). Both inventories also indicate sharp reductions in Hg releases into the environment beginning *ca.* 1982 (materials flow) or 1992 (direct emissions) (Fig. 6.3) that continued through 2000. When compared to peak releases or emissions, the fluxes in 2000 equate to a small fraction (6.1 and 7.5%, respectively). Most of the decline in Hg emissions during this period was related to reductions in emissions from municipal waste combustion (MWC) and medical waste incinerators and, because a large fraction of the emitted Hg from these two sources was likely in the form of GOM, it is likely that these two source types contributed large amounts of Hg to the Everglades during the time frame of peak emissions.

The possibility that the contribution from localized sources historically was much more substantial and significant also is inferred by comparing the degree of enrichment in Hg accumulation rates in the surficial horizon of Everglades sediments to accumulation rates in horizons dated *ca.* 1900. Enrichment ratios averaged 4.0 for WCA-3 and 8.7 for WCA-2 (Rood et al. 1995) compared to typical enrichment ratios ~ 2–4× greater than fluxes that occurred during preindustrial times for relatively pristine background aquatic systems in the northern hemisphere. By comparison, sediment Hg enrichment ratios in Lake Annie, which is a seepage lake[2]

[2] Seepage lakes are defined as having no surface inflows or outflows and typically receive most of their hydrologic inputs from rainfall directly to the lake surface supplemented by water seeping into the lake from the surficial (water table) aquifer. As a result, temporal changes in Hg sediment accumulation rates in seepage lakes are considered a good proxy for temporal changes in atmospheric deposition assuming no watershed disturbances have occurred that otherwise would influence sediment deposition dynamics.

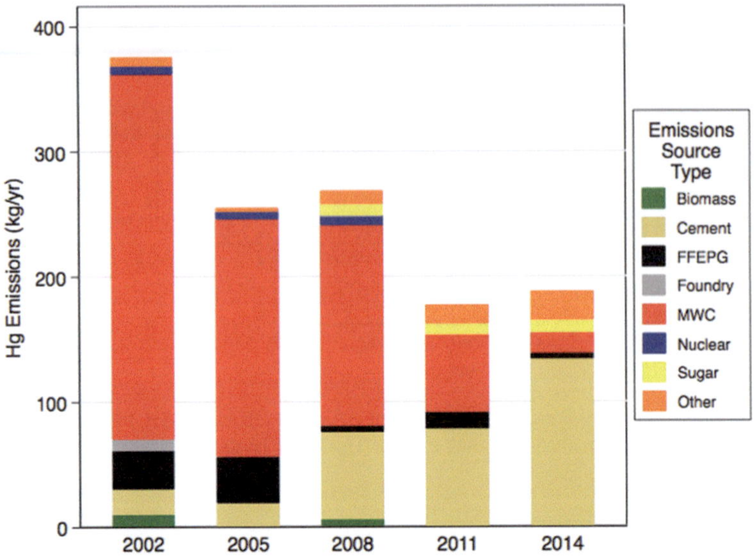

Fig. 6.4 Changes in south Florida (Palm Beach, Broward and Miami-Dade counties) emissions by emissions source category based on NEI estimates published triennially between 2002 and 2014. Each emission category sum includes sources emitting in excess of 4.54 kg/yr (10 lb/yr). FFEPG is fossil fuel electric power generation. "Other" category includes all sources emitting less than 4.54 kg/yr

located north of Lake Okeechobee approximately 203 km from the northern edge of WCA-2 – and because of its location is less likely to reflect any significant impacts from local emission sources – averaged 3.2 ± 0.7 (Engstrom et al., 2003; see Chap. 1, Volume III).

Based on NEI estimates, Hg emissions in the south Florida urban fringe have continued to decline since 2002 (Fig. 6.4) by approximately 50% (19% when compared to the RMB emissions estimate for 2000). These reductions are primarily related to declines in MWC-related emissions, although the reductions have been substantially offset by continued, increasing emissions from cement kiln activity that have corresponded to recent increases in housing construction.

As mentioned previously, local emissions of Hg currently are a quite minor contributor to fluxes of atmospheric Hg deposited in the Everglades and yet wet deposition fluxes in the Everglades are amongst the highest for background sites in the continental US. Why this is so has important policy implications for the restoration of the Everglades and is related to the chemistry of Hg in the free troposphere coupled with the cloud physics of rainfall events in south Florida. The exacerbating effects of these factors on Hg wet deposition fluxes are further reinforced by the frequency and overall volume of rainfall events in the region. This latter fact is evidenced by generally higher annual fluxes of precipitation relative to the rest of the conterminous US other than the northern coastal region of the Gulf of Mexico.

6.3 Gaseous Elemental Hg

From a global perspective, evasion of GEM from terrestrial ecosystems appears to roughly balance inputs of GEM through uptake (Amos et al. 2014, Song et al. 2015; and Horowitz et al. 2017; see Chap. 3, this volume). For the Florida Everglades, losses of GEM to the atmosphere may be relatively even more substantial. GEM losses from the Everglades to the atmosphere include volatilization from the water surface and macrophyte pumping of GEM produced in the bottom sediments as result of reduction of Hg(II). Using flux chambers to monitor GEM evasion from aquatic plants (macrophytes) and the open water surface at an artificial and highly eutrophied wetland system (the Everglades Nutrient Removal [ENR] project), Lindberg et al. (2005) reported that GEM release from macrophytes via transpiration was 20x more important than the GEM flux from the open water, and that the input and release of Hg to the system were roughly balanced at around 70–80 μg Hg m^{-2} yr^{-1}.

It is difficult, however, to scale the Lindberg et al. (2005) results for the ENR to the Everglades Protection Area (EvPA) as a whole for several reasons. First, because of high surface water hydraulic loading rates, surface water inputs of Hg to the ENR exceed direct atmospheric inputs by perhaps a factor of two to or more. This is not the case for the EvPA where, as noted previously, direct atmospheric inputs of Hg exceed surface water inputs by nearly 20:1. Second, as acknowledged by Lindberg et al. (2005), their estimated fluxes are highly uncertain. We have re-examined the Hg mass balance and the uncertainties in the input and output fluxes presented by Lindberg et al. (2005) to evaluate whether their assertion of mass closure is indeed a reasonable conclusion. Our evaluation is based on conducting a Monte Carlo simulation experiment (N = 10,000) where each individual simulation randomly perturbs each input and output flux defined by Lindberg et al. (surface inflow, atmospheric deposition, surface outflow, evasion, and transpiration). Since a mean or median value was not specified, nor a distribution to characterize the uncertainty, we have assumed a uniform distribution within the limits of uncertainty in each flux component. Our analysis indicates that losses – which are dominated by transpiration (83 ± 1%) – exceed inputs on average by 33%, with 25% of the simulations showing excess losses exceeding 53%. Moreover, the mass balance developed for the ENR by Lindberg et al. does not include net burial losses of Hg to the sediments. All these factors suggest that the transpiration rates measured by Lindberg et al. (2005) are not supportable over time (losses greatly exceed inputs) and certainly are not extensible to the EvPA at large where direct atmospheric inputs are by far the predominant source of Hg. Not surprisingly, in their calibration of the Everglades Mercury Cycling Model (E-MCM) to site 3A-15 in WCA-3A as part of the Everglades Pilot Hg TMDL, Harris et al. (2001) predicted macrophyte transpiration rates orders of magnitude lower than the estimates of Lindberg et al. (2005) (0.12 *vs.* 67–80 μg/m^2-yr, respectively). The uncertainty in these estimates, coupled with the implications for biogeochemical cycling of Hg in the Everglades (see Chap. 1, Volume II) are testimony to the need for further research to constrain our understanding of these fluxes.

6.4 Why Atmospheric Deposition Fluxes of Hg to the Everglades Are So High

In general, very high concentrations of GOM are found in the free troposphere. These high concentrations, which can be more than an order of magnitude higher than corresponding concentrations in the underlying boundary layer, are the product of a variety of oxidizing species that react with GEM to produce GOM. These oxidizing species include reactive halogens that are largely marine in origin. Because of south Florida's geographic setting as a peninsula surrounded by marine waters, GOM production rates in the free troposphere are likely higher compared to more inland regions of the continental US. The highly soluble GOM in turn can support high concentrations in Hg in wet deposition when thunderstorm updrafts penetrate upwards through the free troposphere. Convective storms in south Florida are both tall and frequent and, because of their height, can ventilate both the upper as well as lower troposphere and scavenge higher concentrations of GOM.

Hg deposition fluxes in the Everglades thus owe their origins to large-scale processes and their comparatively high magnitude to the subtropical, peninsular geographic setting of the Everglades. While local emissions likely exerted a large impact on Hg deposition in the Everglades that peaked in the late 1980's to early 1990's, reductions in emissions as a result of mandated controls and changes in material use containing Hg have subsequently and largely eliminated the current impact of local sources. As a result, any effective management strategy for mitigating the Everglades Hg problem through reducing atmospheric deposition inputs must rely upon eliminating anthropogenic contributions to the global Hg cycle. As we shall see in Chap. 3 of Volume III, this is problematic both in the immediate short term and extending into at least the next several decades.

References

Amos HM, Jacob DJ, Kocman D, Horowitz HM, Zhang Y, Dutkiewicz S, Horvat M, Corbitt ES, Krabbenhoft DP, Sunderland EM (2014) Global biogeochemical implications of mercury discharges from rivers and sediment burial. Environ Sci Technol 48:9514–9522. https://doi.org/10.1021/es502134t

Atkeson TD, Axelrad D, Pollman CD, Keeler G (2003) Integrating atmospheric mercury deposition and aquatic cycling in the Florida Everglades: An approach for conducting a Total Maximum Daily Load analysis for an atmospherically derived pollutant. Integrated summary. Final report submitted to USEPA. Florida Department of Environmental Protection, Tallahassee

Cohen M, Artz R, Draxler R (2007) NOAA report to Congress: mercury contamination in the Great Lakes. Air Resources Laboratory, Silver Spring. Submitted to Congress on May 14, 2007. 162 pages. Available at: http://www.arl.noaa.gov/documents/reports/NOAA_GL_Hg.pdf

Dvonch JT, Graney JR, Keeler GJ, Stevens RK (1999) Use of elemental tracers to source apportion mercury in South Florida precipitation. Environ Sci and Technol 33:4522–4527

Engstrom DR, Pollman CD, Fitzgerald WF, Balcom PH (2003) Evaluation of recent trends in atmospheric mercury deposition in south Florida from lake sediment records. Final research report. Florida Department of Environmental Protection, Tallahassee

Florida Department of Environmental Protection (FDEP) (2013) Mercury TMDL for the State of Florida. Final report. Watershed Evaluation and TMDL Section. October 24. https://floridadep.gov/sites/default/files/Mercury-TMDL.pdf

Gill GA, Guentzel JL, Landing WM, Pollman CD (1995) Total gaseous mercury measurements in Florida: the FAMS project (1992–1994). Water Air Soil Pollut 80:235–244

Guentzel JL, Landing WM, Gill GA, Pollman CD (1995) Atmospheric deposition of mercury in Florida: the FAMS project (1992-1994). Water Air Soil Pollut 80:393–402

Guentzel JL, Landing WM, Gill GA, Pollman CD (1998) Mercury and major ions in rainfall, throughfall, and foliage from the Florida Everglades. Sci Total Environ 213:43–51

Guentzel JL, Landing WM, Gill GA, Pollman CD (2001) Processes influencing rainfall deposition of mercury in South Florida. Environ Sci Technol 35:863–873

Harris R, Pollman CD, Hutchinson D, Beals D (2001) Florida pilot mercury Total Maximum Daily Load (TMDL) study: application of the Everglades mercury cycling model (E-MCM) to site WCA 3A-15. In: Final report prepared for the United States Environmental Protection Agency and Florida Department of Environmental Protection. Tetra Tech, Inc., Lafayette

Horowitz HM, Jacob DJ, Zhang Y, Dibble TS, Slemr F, Amos HM, Schmidt JA, Corbitt ES, Marais EA, Sunderland EM (2017) A new mechanism for atmospheric mercury redox chemistry: implications for the global mercury budget. Atmos Chem Phys 17:6353–6371. https://doi.org/10.5194/acp-17-6353-2017

Husar JD, Husar RB (2002) Trends of anthropogenic mercury mass flows and emissions in Florida. Final report. Prepared for Florida Department of Environmental Protection under subcontract to Tetra Tech Inc. Lantern Corporation, Clayton

Lindberg SE, Dong WJ, Chanton J, Qualls RG, Meyers T (2005) A mechanism for bimodal emission of gaseous mercury from aquatic macrophytes. Atmos Environ 39(7):1289–1301

Pollman CD, Gill GA, Landing Guentzel WM, Bare D, Porcella D, Zillioux E, Atkeson T (1995) Overview of the Florida Atmospheric Mercury Study (FAMS). Water Air Soil Pollut 80:285–290

RMB Consulting & Research, Inc. (RMB) (1999) South Florida mercury emissions inventory (1996). Final report. Prepared for Florida Department of Environmental Protection. Prepared by RMB Consulting & Research, Inc., Raleigh

RMB Consulting & Research, Inc. (RMB) (2002) Atmospheric mercury emissions from major point sources – Broward, Dade, and Palm Beach counties, 1980–2000. Final report. RMB Consulting & Research, Inc., Raleigh

Rood BE, Gottgens JF, Delfino JJ, Earle CD, Crisman TL (1995) Mercury accumulation trends in Florida Everglades and Savannas marsh flooded soils. Water Air Soil Pollut 80:981–990

Song S, Selin NE, Soerensen AL, Angot H, Artz R, Brooks S, Brunke E-G, Conley G, Dommergue A, Ebinghaus R, Holsen TM, Jaffe DA, Kang S, Kelley P, Luke WT, Magand O, Marumoto K, Pfaffhuber KA, Ren X, Sheu G-R, Slemr F, Warneke T, Weigelt A, Weiss-Penzias P, Wip DC, Zhang Q (2015) Top-down constraints on atmospheric mercury emissions and implications for global biogeochemical cycling. Atmos Chem Phys 15:7103–7125. https://doi.org/10.5194/acp-15-7103-2015

United States Environmental Protection Agency (USEPA) (1996) South Florida Ecosystem Assessment Interim Report. Monitoring for adaptive management: implications for ecosystem restoration. Region 4 SESD and ORD. EPA-904-96-008. December 1996

United States Environmental Protection Agency (USEPA) (2005) Technical support document. Revision of December 2000 regulatory finding on the emissions of hazardous air pollutants from electric utility steam generating units and the removal of coal and oil-fired electric utility steam generating units from the Section 112(c) List: Reconsideration, U.S. Environmental Protection Agency, Washington, DC, October 21, 2005